Construction Estimates

Contents

Preface   xi
Abbreviations   xiii

Chapter 1. General Methods and Procedures   1

Preliminary Work   2
The Specifications   3
  General conditions   4
  Alternates   4
  Allowances   4
Before Taking Off   7
Order of Taking Off   9
Time-Saving Practices   11
  Mathematical shortcuts   12
  Collection sheets   14
Summary of General Rules for Taking Off   16

Chapter 2. Excavation and Site Work   21

Visiting the Site   22
  Site clearing   23
  Stripping topsoil   25
Building Excavation: Example 1   25
  The take-off (W.D. 2.1)   26
  Notes on the take-off (W.D. 2.1)   27
Building Excavation: Example 2   27
  The take-off (W.D. 3.1)   28
  Notes on the take-off (W.D. 3.1)   28
Rock Excavation   29
  The take-off (W.D. 2.1)   30
  Notes on the take-off (W.D. 2.1)   30
  Sheeting for excavations   31
Site Grading   32
  The take-off (W.D. 2.2)   35
  Notes on the take-off (W.D. 2.2)   36
  Miscellaneous landscape items   36
Roads and Pavings   36
  The take-off (W.D. 2.3 and 2.4)   38
  Notes on the take-off (W.D. 2.3 and 2.4)   40

Utilities	40
The take-off (W.D. 2.5)	42
Notes on the take-off (W.D. 2.5)	43
Utilities: Septic Tanks and Distribution Boxes	44
The take-off (W.D. 2.6 and 2.7)	44
Notes on the take-off (W.D. 2.6 and 2.7)	47
Miscellaneous Utilities Items	49
Sundry site work items	50
Site Work: Deep Pit in Water	51
The take-off (W.D. 2.8)	52
Notes on the take-off (W.D. 2.8)	53

Chapter 3. Concrete   55

Wall Perimeter	55
Concrete Foundations	57
The take-off (W.D. 3.1)	59
Notes on the take-off (W.D. 3.1)	60
Areaway and Stairs	61
The take-off	62
Notes on the take-off	63
Equipment Pads	64
Take-off of equipment pads	64
Notes on the take-off	64
Take-off of concrete floor pipe trenches (W.D. 3.2)	65
Notes on the take-off (W.D. 3.2)	66
Take-off of concrete retaining wall in W.D. 3.3	67
Notes on the take-off (W.D. 3.3)	67
Concrete Columns	67
The take-off (W.D. 3.4)	68
Notes on the take-off (W.D. 3.4)	68
Concrete Beams and Suspended Slabs	69
The take-off (W.D. 3.5)	70
Notes on the take-off (W.D. 3.5)	73
Concrete canopy	74
Notes on the take-off	75
Concrete Interior Stairs	75
Notes on the take-off	76
Miscellaneous Formwork Items	76
Miscellaneous Concrete Items	78
Sundry Items Taken Off with Concrete	82
Metal Sundries for Concrete Work	84

Chapter 4. Reinforcing Steel   85

Take-Off for Foundation (W.D. 3.1)	87
Notes on the take-off (W.D. 3.1)	88
Take-Off for Retaining Wall (W.D. 3.3)	88
Notes on the take-off (W.D. 3.3)	88
Take-Off for Columns (W.D. 3.4)	89
Take-Off of Steel for Suspended Slab (W.D. 3.5)	90
Notes on the take-off (W.D. 3.5)	92

Chapter 5. Structural Steel — 93

- Take-Off for Structural Steel (W.D. 5.1) — 94
  - Notes on the take-off (W.D. 5.1) — 95
  - Miscellaneous iron — 96

Chapter 6. Masonry — 97

- Preliminary Analysis — 98
- Estimating Brick Quantities — 99
  - Concrete block backup — 103
- Single-Story Building — 104
  - The take-off (W.D. 6.1) — 104
  - Notes on the take-off (W.D. 6.1) — 105
  - Common-brick backup — 106
- Four-Story Building — 107
  - The take-off (W.D. 6.2) — 109
  - Notes on the take-off (W.D. 6.2) — 110
- Exterior Walls with Preformed Waterproofing — 112
- Interior Masonry Partitions — 113
  - Working drawings 6.3 and 6.4 — 114
  - The take-off (W.D. 6.3, 6.4) — 117
  - Notes on the take-off (W.D. 6.3, 6.4) — 119
- Masonry—Miscellaneous Items — 120
- Stonework — 122
  - The take-off (W.D. 6.2, 6.5) — 123
  - Notes on the take-off (W.D. 6.2, 6.5) — 124

Chapter 7. Carpentry — 127

- Exterior Carpentry — 127
  - The take-off (W.D. 6.2, 7.1) — 128
  - Notes on the take-off (W.D. 6.2, 7.1) — 130
- A Classroom — 133
  - The take-off (W.D. 7.2, 7.3) — 133
  - Notes on the take-off (W.D. 7.2, 7.3) — 135
- Wood-Framed Buildings — 136
  - The exterior — 137
  - The take-off (exterior—W.D. 7.4–7.9) — 138
  - Notes on the take-off (exterior—W.D. 7.4–7.9) — 142
  - The interior — 143
  - The take-off (interior—W.D. 7.4, 7.5) — 144
  - Notes on the take-off (interior—W.D. 7.4, 7.5) — 147
- Miscellaneous Rough Carpentry — 148
- Miscellaneous Finish Carpentry — 149

Chapter 8. Alteration Work — 151

- Underpinning — 152
  - The take-off (W.D. 8.1) — 153
  - Notes on the take-off (W.D. 8.1) — 153
- Excavation and Concrete for Piping Work — 154
  - The take-off (W.D. 8.2) — 154
  - Notes on the take-off (W.D. 8.2) — 155

Removing Walls (Shoring)	156
The take-off (W.D. 8.3, 8.4)	156
Notes on the take-off (W.D. 8.3, 8.4)	159
Remodeling	159
The take-off (W.D. 8.5)	160
Notes on the take-off (W.D. 8.5)	162
Miscellaneous Items	163
Demolition	163
Shoring	163
Partitions and interior walls	163
Roofs adjoining new roofs	163
Occupied buildings	163

Chapter 9. Job Overhead — 165

Bonds and Insurances	165
Bid bonds	166
Permit bonds	166
Insurances	166
Builder's risk insurance	166
Health and welfare funds	167
Pension funds	167
Government payroll taxes	167
Field Overhead	167
Superintendent	167
Supervisors	167
Engineering	168
Timekeepers and material clerks	168
Watchmen	168
Job offices and shanties	168
Trucking	168
Temporary light and power	169
Temporary water	169
Temporary heat	169
Other Overhead Items	169
Permits	169
Equipment	169
Small tools	169
Clean-up	170
Glass cleaning	170
Glass breakage	170
Winter protection	170
Signs	170
Progress photographs	170
Premium time (overtime)	171
Labor wage increases	171
Travel time	171
Subsistence allowance	171
Overhead in general	172

Chapter 10. Subcontractors' Bids — 173

Heating, Ventilation, and Air Conditioning	174
Electrical	174
Structural Steel	174
Miscellaneous Steel and Iron	174

Roofing and Flashings	175
Glass and Glazing	175
Millwork	175
Miscellaneous Items	175
Subbids in General	176
Plumbing	176
Metal Windows and Curtain Walls	176
Doors and Door Frames	177

Chapter 11. Industrial Building — 179

General Contractor's Estimate	179
Outline specifications	179
Take-Off Industrial Building	184
Working drawings 9.1 to 9.4	184
Notes on the take-off (W.D. 9.1 to 9.4)	199
Excavation	199
Concrete	200
Reinforcing steel	200
Masonry	200
Carpentry	201
Alternates	201

Chapter 12. Industrial Building—The Estimate — 203

Alternates	203
Notes on the Estimate	212
Excavation and site	213
Formwork	213
Concrete	214
Masonry	214
Carpentry	214
Job overhead	215
Subcontractor bids	215
Mail	216
Summary sheet	216
Alternate bids	216

Chapter 13. Pricing the Estimate — 219

The Good Estimator	220
Becoming a good estimator	220
Examples of Unit Prices	221
General remarks on compiling unit prices	221

Appendix. Mensuration and Tables — 229

Index 235

Preface to the Third Edition

There are books on almost every aspect of building construction, including many that deal with pricing the estimate. First published in 1961, this is still the only book devoted to the daily work of the construction estimator—one that shows how he extracts the quantities from the drawings, compiles and prices the estimate, and prepares the bid. This book, then, could be called "The Estimator at Work."

Here, although the general approach to estimating has not been overlooked, the emphasis is on showing it in action. Examples are included to show how quantities are taken off for all of the trades customarily handled by the general contractor, each with its particular working drawing. In order that the reader may keep the appropriate drawing before him while he follows each take-off, item by item, these have been bound into a separate booklet for which a pocket in the back cover has been provided.

The examples show a methodical system of taking-off. Many shortcuts and time-saving devices are included, showing how computations may be simplified and initial calculations made to do extra duty by being retained for reuse at several stages of the take-off.

The chapters on alterations, overhead, and subbids—usually given scant attention in estimating books—have been treated rather fully, in recognition of their importance to all who work on estimates.

The third edition contains all of the take-off examples and general text of the first and second editions—plus expanded treatment of the general estimating procedures, pricing of general conditions and overhead, new take-off techniques for masonry, and most important the complete drawings, specification data, take-off and priced estimate of a $660,000 building. The pricing has been updated to reflect current prices.

To round out the presentation, there is a chapter giving full details of many unit prices, showing how the prices of labor and material are compiled.

This book, then, is the estimator at work, from the opening of the drawings to the closing of the bid.

The aim throughout has been to show you how to save time, work, and money while achieving greater accuracy than is yielded by less systematic estimating methods. You may be in a general contractor's office; in the masonry, carpentry, excavation, or steel business; or you may be preparing budget estimates for architects, engineers, or building owners. Whatever your concerns or your interests, if they are connected with construction estimating, this book was written for you.

Acknowledgments

The author wishes to thank The Architects Collaborative (Cambridge, Mass.); Kilham, Hopkins, Greeley & Brodie, Architects (Boston, Mass.); and Korslund, LeNormand & Quann, Architects (Norwood, Mass.) for permission to use certain details and drawings. Thanks are also extended to Mr. Anthony T. Conti and Mr. Vincent Iannone who made the finish drawings from the author's rough sketches, and to Frances Murtaugh for revising the manuscript and patiently dealing with the three coauthors concurrently.

Norman Foster
Theordore J. Trauner, Jr.
Rocco R. Vespe
William M. Chapman

Abbreviations

access.	accessories	conc.	concrete
alum.	aluminum	cont.	continuous
arch.	architect	C. P.	concrete pipe
av.	average	cupb.	cupboard
B.	bottom	CY	cubic yards
bast.	basement	D. H.	double hung
bd.	board	dia.	diameter
bdg.	boarding	dist. box	distribution box
BF	board feet	dr.	door
bit.	bituminous	drg.	drawing
bldg.	building	D. U.	dwelling units
bm.	beam	DW	dry wall
B. P.	base plate	ea.	each
b. s. m.	both sides measured	e. f.	each face
C	100	e. w.	each way
c.	courses (on masonry drawing)	exc.	excavation
carbo.	carborundum	exp.	expansion
C. B.	catch basin (on utilities drawing)	ext.	exterior
C.B.	chalkboard (on carpentry drawing)	fcg.	facing
c-c	center to center	Fig.	figure
cem	cement	FL	furring & lathing
Cem Fin	cement finisher	flr.	floor
ceram.	ceramic	fndtn.	foundation
cert.	certified	fr.	frame
CF	cubic feet	frmg.	framing
Chan	channel	F. T.	facing tile
C. I.	cast iron	ftg.	footing
CJ	control joint	fus	fusible
CLF	100 lineal feet	galv.	galvanized
clg.	ceiling	gen.	general
C. of W.	clerk of works	G. F. T.	glazed facing tile
col.	column	gl.	glazed
com.	common	G1F	glazed one face
comb.	combination	G2F	glazed two faces
comm.	commercial	G1S	good one side

Abbreviations

gr.	grade	plmg.	plumbing
grano.	garnolithic	ply.	plywood
G. W. L.	ground water level	porc.	porcelain
gyp.	gypsum	pr.	pair
H.	header (on masonry drawing)	psf	pounds per square foot
H. B.	heavy bending	psi	pounds per square inch
HC	hollow core	pt.	paint
H. D. C. I.	heavy-duty cast iron	ptn.	partition
hgt.	height	PVC	polyvinyl chloride
H. M.	hollow metal	QT	quarry tile
hor.	horizontally	R.	risers (as in "14R")
HVAC	heating, ventilating, & air conditioning	rad.	radius
H. W. L.	heavyweight Lally	R. C. P.	reinforced concrete pipe
I. D.	internal diameter	reinf.	reinforcing
ins.	insurance	retg.	retaining
int.	interior	rm.	room
inv.	invert	R. T.	rubber tile
jnt.	joint	S.	stretcher (on masonry drawing)
jntd.	jointed	san.	sanitary
jst.	joist	SC	solid core
Kal.	kalamein	S. E.	square edged
kit.	kitchen	sect.	section
L. B.	light bending	SF	square feet
lb	pounds	span.	spandrel
LF	lineal feet	sqr.	square
lg.	long	st.	station
lino.	linoleum	str.	straight
L. S.	lump sum	susp.	suspended
M	thousand	SY	square yards
magn.	magnesium	T	ton
mat'l	material	T.	top
max.	maximum	T.B.	tackboard
M.H.	manhole	T. C.	terra cotta
min.	minimum	temp.	temporary
NIC	not included	T. & G.	tongued and grooved
op	operator	typ.	typical
opg.	opening	V. C.	vitrified clay
PCC	precast concrete	vert. (or V.)	vertically
pcs.	pieces	W. D.	working drawing
perim.	perimeter	W. I.	wrought iron
Pl.	plaster	wind.	window
pl.	plate		

Chapter

1

General Methods and Procedures

A construction estimate is developed in three separate but interlocking parts, take-off, pricing, and subbids. Developing an estimate is an exacting and demanding professional task, but it is not an exact science. A good estimator is a good construction technician who, through experience and special interest, has the ability to measure quantities accurately and to gauge costs. A good estimate is built around a good set of quantities and a proper feeling for cost, rather than being a by-product of statistics.

A competitive bid is a one-shot deal, a total commitment from which there is normally no withdrawal. Even a budget estimate may be the basis for commitment to funding, land purchase, bond issues, and loan agreements. A constant awareness of the far-reaching importance of the figures and final price is a vital part of good estimating. Sound techniques, conscientious application, imagination, a sensitivity for cost, and the realization of the seriousness of the responsibilities—these are the essence of good estimating. Casualness, slackness, guesswork, and blind gambling—these make up the unsafe, irresponsible, poor estimate.

Starting from scratch, in a matter of days the estimator must produce in theory what the entire field force will produce only after many weeks or months of actual work. The success or failure of a construction company begins with the quality of its estimating. Other factors are vital to a company, but they follow after and are very much dependent on the prices a firm gets for the work it does. Job planning, supervision, and purchasing can affect the degree of financial success, but the estimate is the controlling base.

Even luck plays a part, but it should never assume more than a minor supporting role. Bad luck, or what is usually called bad luck, is more often than not merely bad judgment. Estimating and drive are the heart of success in the construction industry.

Every construction project is different from every other project; each estimate is an entity, the complete and particular summary of the best available cost information pertaining to a specific bid. It is separate and different from all previous estimates and must be treated as such. Experience, cost records, knowledge of the area, and all such factors equip the estimator, but a particular project being bid has factors peculiar to itself and can never be a carbon copy of any previous project or bid.

A good estimator knows the skills, strengths, and weaknesses of his or her company and through training, perception, and experience molds them to suit the particular project being bid.

PRELIMINARY WORK

Before starting to take off the quantities for an estimate, you must attend to some preliminary chores in order for the estimate to go together with the minimum of difficulty. Your participation as a bidder must be publicized, and you must acquire a general familiarity with the project to be bid.

Having obtained the bidding documents, you should check the specifications for date and time of bidding, addenda, bid security required, mandatory meetings, subtrades, and materials for which prices must be solicited. As a first step, that is, before preparing the "Specification take-off" which is discussed below, you should ask the construction reporting service to include your firm as a general contractor bidder in their news service listing for the project. The longer your name is carried in the reports, the better are your chances for getting quotations from suppliers and subcontractors. You should include your name or the name of the contact person to make it easier for those having questions about the bid. F. W. Dodge Division of McGraw-Hill, Inc., issues a daily construction news bulletin in most major U.S. cities. Also check for local construction reports in your area.

At this stage you should also notify your bonding agent that you are bidding the project and give him or her all the needed data, such as name of job, owner, bid date, approximate value of the project, bid security required, and form of security required (whether bid bond, certified check, or whatever). A construction performance and payment bond is the owner's protective insurance. Under the bond, the surety company provides a monetary guarantee concerning the completion of the contract or the payment of all labor and material bills relative to the contract, should the contractor default. The bonding company does not relieve the contractor of responsibility; it backs the contractor up and expresses its faith in the contractor by writing the bond on his or her contracts.

One further preliminary step should be taken at this early stage of preparing an estimate. Requests for subbids and material prices should be sent out. This is usually done on preprinted "bid request" postcards. The format of the card is simple. It should identify the job, architect, owner, and date bids are requested. The card should state something like "We would be pleased to receive your bid for work in your line for the above-named project." The cards should be mailed to all probable subcontractors and suppliers. A computerized mailing list of subcontractors, suppliers, and specialty manufacturers should be maintained to simplify this process.

These bid request cards should not be sent out until you have made the "Specifications take-off" described below. The specifications take-off will provide the list of trades and materials for which prices should be solicited, thus saving a duplication of effort.

THE SPECIFICATIONS

Before beginning to take off quantities, the specifications should be read and all the factors vital to the estimate noted. The best way to do this is to make a "Specifications Take-Off" on plain ruled paper, such as the standard $8\frac{1}{2}$ by 11 pads. The goal should be to get information pertinent to the estimate in a brief, easy-to-read-and-check form to ensure that you miss nothing and to save the time which would be lost by continually referring to the specifications during the preparation of the estimate.

Reading the specifications need not be a laborious job. You will find as you become more experienced that although there are certain items to look for in particular, specifications may contain many standard paragraphs that you are familiar with and need only scan. It is important, however, to always check the clauses that deal with changes, time extensions, and disputes. It is not necessary to read the subcontractors' sections of the specifications, except to ascertain what items are included, or perhaps to see what reference is made to "work by others" (particularly when "others" refers to the general contractor).

A specifications take-off lists the name of the project, the architect, the owner, the date bids close, and the amount of the bid bond or other bid security. The general conditions are read and all salient factors noted such as completion time, liquidated damages, changes clauses, disputes, time extensions, payment terms, retainage, scope of the work, items not in the contract, items by others, allowances, and alternates. Special care should be taken to check the method of payment for items such as excavation, concrete, and shoring. It is not uncommon for these items to be paid based on plan quantities and not actual installed quantities. The specifications take-off should be concise, using abbreviations wherever the meaning will not be lost to you.

In the detailed chapters of the specifications, however, all sections that are general contractor's work must be read carefully, and the various items and materials that are required must be noted. Subcontractors' sections should be scanned, and any special inclusions or exclusions noted. It is especially important to check the mechanical trades for items such as temporary heat, storm and sanitary sewers, water service, duct lines, cutting and patching, layout, sleeves, and concrete pads for mechanical items. It may be necessary to check other sections of the specifications against the mechanical-trade sections to determine which of these items are to be provided by the general contractor.

Subtrade sections that contain "supply only" items should also be read carefully, to ascertain the extent of the general contractor's responsibility for unloading, protection and setting.

A typical specifications take-off follows (boldface type indicates subitems):

Owner.—Mass. School District
South Elementary School   Arch.—Nodder & Pulski
Baysville, Mass.   Bids—June 9, 199x, 2 P.M.
Completion—340 cal. days   Cert. check—$18,000
L.D.'s—$500 per day   Payments—monthly

General conditions

By general contractor: permits, progress photos, temporary heat (fuel only), trailer for owner's representative, 2 phones, sanitary facilities (portable toilet).

Not in contract: fire ins. on bldg., utilities beyond site boundary, furniture, drapes.

Generally no overtime work, 10 percent retained 90 days, 1-year maintenance.

Alternates

1. Add for roof changes (see Drg. 12A)
2. Deduct ceramic tile & add resilient flooring—Rms. 1A, 3, 9, & 27.

Allowances

Finish hardware	$ 12,500	(supply only)
Stage curtains	$ 6,100	(installed)
Kitchen equipment	$36,200	(installed)

Section 1.   Excavation & site work

Clear site of trees, etc., & grub up roots.
Strip & stockpile loam. Pumping.
Bldg. excavation. Sheathing.
Backfill—clean bank gravel 3 ft around all walls.
12 in. bank gravel under ground slabs, platforms, & walks.
Site grading—6 in. low for lawns; 15 in. low for roads; 16 in. low for walks:
Excavate utilities—storm sanitary, water, telephone, fire, oil tank, oil line, electric.

Storm drain—R.C.C. pipe; catch basins, head walls.

Sanitary sewer—exc.; catch basins & manholes only (plumber does piping).

Concrete envelope (2,500 psi) for electric service.

Water service (street to bldg.).

Pavings—12½ in. graded gravel + 2½ in. bitumen + bit. curbs.

Walks—12 in. bank gravel + 4 in. concrete (3,000 psi)+ 6-in. × 6-in. × 10/10 mesh. Broom finish.

½-in. × 4-in. expansion jointing on 30-ft centers.

Flagpole base.

Rebuild stone wall at west boundary.

Lawns & planting.

Section 2. Concrete

2,500-psi concrete—footings, foundation walls, ground slabs.

3,000-psi—all other.

Reinforcing steel.

Mesh—6 in. × 6 in. × 6/6 for ground slabs.

Rub exposed walls, ceilings, etc.

Trowel floors + 2 coats liquid hardener for exposed concrete floors.

Grano. floor topping 1:1:2 as drgs.+color admix.

Stair fill (3,000 psi) + abrasive grits ¼ lb per SF.

Entrance platforms and steps—float + abrasive ¼ lb per SF.

Lightweight concrete roof fill—1:7 (min. 2 in.)

Set col. base plates, bolts, etc.

Concrete tests (4 cylinders per day of pour or 50 cy).

Waterproofing admix for boiler rm. floor & walls.

Section 3. Structural steel (shop paint + field paint 1 coat)

Bar joists.

Section 4. Masonry

Ext. face brick—waterstruck; Flemish header in 6th course? Backparge.

Common brick—backup at beam bearings.

Concrete block—backup for gym. walls.

Cinder block—backup & ptns.

Gypsum block ptns.

Limestone + dowels & anchors.

Block reinforcing mesh—all backup block & ptns., continuous at every 2d course.

Facing brick (5 × 12) clear glazed + plain trim shade caps as drgs. Bullnosed jambs and quoins. Coved base in toilets only.

Brick anchors & straps.

Mortar—*1:2:6* (ext. walls + waterproofing).

16-oz. copper dam at expansion joints.

Section 5.   Miscellaneous metal (check specs. for items)

Hollow metal doors & frames
Kalamein doors.

Section 6.   Metal toilet ptns. & screens

Reinforced panels for handicap grab bars and accessories

Section 7.   Acoustical tile

Section 8.   Gypsum roof + formboard + subpurlins

Section 9.   Fire extinguishers

Section 10.   Furring, lathing, & plastering

Section 11.   Carpentry

Roof nailers & eaves blocking (pressure treated).

Framing canopy + 2-in. T. & G. deck.

6-in. batt insulation at eaves.

Cemesto panels ($1\frac{1}{4}$-in. & $1\frac{1}{16}$-in. at window walls).

1 × 3 furring on 16-in. centers—ext. walls for plaster.

1 × 2 furring on 12-in. centers—clg. for acoustic tile.

Chalkboards, tackboards, & wood paneling.

2 × 3 studding at wardrobes and closets.

Framing & 1-in. T. & G. deck—catwalks.

Blocking for cabinets & casework.

Rough bucks and window blocking (treated).

Frmg. stage floor + 1-in. S.E. subfloor (diagonally).

Millwork.

Wood doors & frames.

Wood windows.

Overhead doors.

Section 12. Glass & glazing

Section 13. Painting

Section 14. Terrazzo & ceramic tile

Section 15. Resilient floors (VCT & sheetgoods)   **Rubber core base, 4 in.**

Section 16. Plumbing (includes sanitary sewers except excavation & manholes)

Section 17. Heating & vent. (includes labor & set units for temp. heat)

Section 18. Electrical (includes all temporary wiring: gen. contractor pays for power)

Subitems (indicated by boldface type in the list above) should be highlighted or double-underlined in blue. Use of a red pen or pencil should be restricted to deductions and queries. The reason for highlighting or underlining the subitems in colored pencil on the specifications take-off is to make them stand out when you write up the subsheet of the estimate and when you send out the bid request cards.

A final check should be performed to make sure each item shown on the specifications table of contents has been addressed in the specifications take-off sheets. If there is any doubt about the intent of the specifications, a small question mark should be made next to the item in question (see, for example, the face-brick bond item in the sample specifications take-off). The specifications can be rechecked when that item is being taken off. Except for resolving unclear items, the specifications can now be put aside, and only the specifications take-off used.

The purposes of a specifications take-off are to give you an overall knowledge of the items before you start taking them off individually, to determine the allocation of items between the general contractor and the subcontractors, to subdivide sections that include more than one trade, and to save you time in taking off by providing a convenient summary of the items required. The specifications take-off will also provide the list of trades for the subcontract summary sheet and the bid request card process.

BEFORE TAKING OFF

Every construction estimate is based on a quantity survey, sometimes called the take-off. The quantity survey is an extraction from the drawings and specifications of all the labor and material required for the project. No bid can be more accurate than the quantity survey on which the estimate is based. A quantity survey, properly made, is much more than simply a list of so many cubic yards of concrete, so many bricks, so much of this, and so much of that. A good take-off shows everything necessary to prepare a proper estimate for the job—not for any job, but for the particular job that is being bid. An office engineer-estimator must be able to work accurately, quickly, and methodically in taking off. There are many "tricks of the trade" that will save time, reduce errors, and improve accuracy; there are no shortcuts, however, that can be taken at the expense of accuracy. "Near enough" is not sufficient. Fractional figures may be rounded off for some items, but such adjustment must be judicious and controlled. For example, consider a concrete item that is 120 ft 9 in. $\times$ 1 ft $3\frac{1}{2}$ in. $\times$ 11 ft 10 in. This item cannot be altered to 121 ft $\times$ 1 ft 4 in. $\times$ 12 ft just because those figures would be easier to work with—an error would result of almost $3\frac{1}{2}$ CY. However, having properly computed the item in cubic feet, the conversion to cubic yards may be adjusted to the nearest half cubic yard. This principle of control in rounding off quantities will be evident in the examples shown throughout this book.

Before starting to take off quantities, examine the drawings—all of them. Take a quick look through the entire set of drawings for an idea of the layout, type of building, number of floors and general design, and also the order of the drawings. This quick look takes only a few minutes, yet many little things will stick in your mind and save you time later.

There are three basic rules for taking off quantities:

1. Measure everything as it shows.
2. Take off everything that you can see.
3. If it is different, keep it separate.

1. The first rule—"measure everything as it shows"—simply means take it off exactly as it shows on the drawings; do not "approximate," do not "average," do not "round it off," and do not change something because you think you know better than the architect or engineer. A measurement of 21 ft 9 in. is not 22 ft. If calculations involving odd inches are beyond your mathematical ability, then estimating is not the work for you. You want to save time? There are ways—many ways—to save hours of your take-off time; they are evident throughout this book. But as for changing something because you think you know what is intended, remember that your take-off is to be used for an estimate on a building "as per plans and specifications"—not for what you imagine is required. If there is a discrepancy between details or between drawings and specifications, you may have to ask the architect or engineer for clarification; but never take off what is not called for. Do not attempt to superimpose your own construction knowledge on the architect's drawings, for you are bidding only on what is shown. If changes are made subsequent to the award of a contract, they are a matter for adjustment between the contractor and the owner.

2. The second rule—"take off everything that you can see"—means take it all off; do not deliberately leave out anything. You will undoubtedly make

some mistakes; we all do. You cannot hope to break down a building into hundreds of small parts and not miss something or make some kind of error. But in order to reduce the probability of error—and so minimize the risk inherent in any bid—you must take off everything that you can see.

If you follow these two rules—taking off everything that you see and taking it off exactly as it shows—you should have a good take-off as least as far as quantities go.

3. The third rule—"if it is different, keep it separate"—means that you must separate items that will require special consideration, and that you must not mix items that will require different unit prices. Ordinary foundation walls should be carried separately from walls "up in the air" (concrete work which will involve a crane or pump for pouring and may also entail more costly formwork). Curved or exposed concrete walls usually require expensive formwork; keep these items separate. Special face-brick patterns, brick arches, and herringbone brick paving are all examples of items that are abnormally expensive and should be kept separate. Machine trench excavation costs more than machine bulk excavation; these items should not be lumped together. Work that will require more than normal consideration in pricing the estimate is often best handled by taking off extra cost items. The quantities must fully describe and measure all the work involved. For example, formwork for foundation walls may be taken off as a main item—"Exterior foundation walls—14,340 SF"—plus separate extra cost items: "Formwork 4-in. brick shelf—1,620 SF; Wall pilasters average 1–6 × 8–0 × 7–3—25 each; 1-in. chamfer strip to walls—386 LF." If there is something very unusual about an item, then it should be noted: "See Drg...." or "Bottom 2 ft below G.W.L.," or whatever is applicable. Pricing an estimate is always a difficult and problematical matter, but a good take-off will reduce the difficulties.

If the take-off is correct and complete, the estimate can be priced with confidence. A poor take-off, a rough, half-guessed take-off, or a take-off that does not reflect the special requirements of the job concerned, will yield an unsatisfactory estimate. Errors in the field can be costly, but they can be found and corrected. Errors in the take-off, however, have to be absorbed by your employer. There is no adjustment possible for estimating errors once the bid has been submitted. The estimating department always works under pressure. There is usually no time for checking back and forth; it is not usually possible to do the take-off twice in order to be sure. A certain amount of checking of extensions must be done, but the accuracy of the take-off itself cannot be checked. Contractors involved in competitive bidding cannot afford to double their estimating staffs in order to take off each job twice. The office engineer-estimator taking off the job must know exactly what he or she is doing, must work quickly and methodically, and must have self-confidence.

ORDER OF TAKING OFF

There are several orderly ways for a take-off to progress. One cannot be dogmatic and say that a particular trade must be taken off first for every job.

Many projects must be approached quite differently from the normal way. Such jobs excepted, however, the take-off should proceed in a definite pattern—not only trade by trade, but also item by item within each trade. Excavation should not be taken off first. The excavation items depend on and follow the below-ground structure, so the substructure items should be taken off before the excavation. That way you will know just what is involved when you start taking off the excavation items, and you will have many needed quantities and measurements already computed. The suggested order of take-off is:

1. Concrete   Substructure

                      Superstructure

                      Finishes

2. Masonry   Exterior

                      Interior

3. Carpentry   Rough

                      Finish

4. Finishing trades   Any that the general contractor must do

5. Excavation   Building

                      Site

6. Site work

7. Alternates

With the concrete (or even just the substructure concrete) items completed, it would be perfectly in order for you to turn to the excavation and take off either the building excavation or the entire excavation and site work before proceeding with the masonry. I prefer the order indicated above because I like to work through the building methodically, trade by trade, before starting on the excavation and site work. This preference is not just a personal prejudice. Very often during the process of taking off, an opportunity will occur to visit the site, and a preliminary look at the site will often affect the approach to the excavation items. There may even be specific considerations that necessitate a site visit before taking off the excavation—considerations such as the depth and quality of topsoil, the water condition, street and curb conditions, trees, and access. For a job that is heavy on excavation or site work, making two visits to the site—one before taking off the excavation items and one before pricing the estimate—is good practice.

The order of taking off should not vary except for very special jobs. The purpose of establishing and keeping a particular order is to so order your mind that one item will follow another in a methodical pattern not only from trade to trade but through the entire take-off. Consider a specific trade—say, concrete. The take-off should progress through the building: pier footings, foundation piers, wall footings, foundation walls, entrances, steps, platforms, ground slabs, and other substructure items; columns, beams, suspended slabs (first floor, second, and so on), roof fill, and other superstructure concrete; floor finishes, rubbing, curbs, bases, and sundries. A similar order of take-off should be developed for each trade. Having developed an order of taking off, stay with it—use it always. Every item will have its place, the take-off will go smoothly and swiftly, and the probability of missing an item will be reduced to a minimum; also, a particular item will be easily found when you refer back to the take-off sheets to check something.

TIME-SAVING PRACTICES

Many "tricks of the trade" that save time will be shown in the take-off examples that follow, but certain basic rules may be stressed now.

1. Never use a long word if a short word will do.
2. Abbreviate words whenever possible. "Exc.," "Conc.," "Forms.," "San. Sewer," and "tel." are simple examples. The examples of take-off sheets throughout this book will use many such abbreviations. (A list of abbreviations is given at the beginning of the book.)
3. Keep all dimensions, figures, and areas that might be useful later. A prime example of such a figure is the total exterior-wall perimeter. This figure, first obtained when taking off the concrete walls, could be needed for damp-proofing, rubbing concrete, concrete footings, exterior masonry, plates, stone bands, copings, and eaves items.
4. Learn the "27 times" multiplication table (Table 1.1). It is a simple table, quickly learned, and can be used constantly in converting concrete and excavation items from cubic feet to cubic yards.

Always be ready to take advantage of combinations of figures that lend themselves to speedy reckoning. The "27 times" table is often especially useful. For example,

$$94\text{--}0 \times 1\text{--}0 \times 13\text{--}6 = 47 \text{ CY}$$

13–6 being half of 27–0, the computation can be reduced to 94 divided by 2. Or, in figuring the excavation of wall trenches,

$$216\text{--}0 \times 4\text{--}6 \times 4\text{--}0 = 144 \text{ CY}$$

216 is a multiple of 27 (8), so to convert to cubic yards, the calculation becomes 8 times $4\frac{1}{2}$ times 4, or 8 times 18.

Another example, for 2×8 joists:

$$30 \times 14 \text{ LF} = 560 \text{ BF}$$

Since 2×8 requires $1\frac{1}{3}$ BF per LF, the calculation is actually $30 \times 14 \times 1\frac{1}{3}$; it is simplest, however, to multiply 30 by $1\frac{1}{3}$ first, giving 40, and then multiply 40 by 14 (560).

TABLE 1.1 27 Times Table

1 × 27	=	27
2 × 27	=	54
3 × 27	=	81
4 × 27	=	108
5 × 27	=	135
6 × 27	=	162
7 × 27	=	189
8 × 27	=	216
9 × 27	=	243

Mathematical shortcuts

There are many aids to speedy calculating, as will be seen in the take-off sections of this book. Reference to the sample take-off sheets will show that all dimensions for building items throughout this book are written in feet and inches, not in decimals. Even if you intend to use a calculator, it is best to list the figures in the units that are used on the drawings. Architectural and structural drawings are usually dimensioned in feet and inches. If one is using a calculator, the conversion to decimals can be made when putting the figures into the calculator. For those who wish to compute the quantities without using a calculator or do not have a calculator at hand, feet and inches are as easy to work with as decimals. The disadvantage of changing the figured dimension into decimals is that the more conversions the more steps, and the more steps the greater the probability of error. Every step saved when taking off must reduce the probability of error. For each of us, there is a certain probability of error—whether an average of one error per 10 calculations or one per 1,000; thus if we reduce the number of operations, we also reduce the number of errors.

Estimators usually use a calculator for converting their quantities and pricing the estimate. There are even a few mathematically minded purists who prefer to do all their computations without using a calculator. Whether you are calculating manually by necessity or choice, however, the quickest and simplest method is by duodecimals or ratio, or (more often than not) a combination of these two methods, which are explained in the Appendix.

The basic rule for multiplication is to multiply the whole numbers and the largest factors first, before multiplying small fractional parts, so the error involved in rounding off fractions or decimals is reduced to a minimum. Given three dimensions to produce a cubic quantity, two of them small figures and one large, if the two small figures were multiplied first and a fraction or a couple of decimal points dropped, then the error would be multiplied by the amount of the large figure.

Consider the following calculation for concrete foundation walls:

General Methods and Procedures

Concrete foundation walls

By feet and inches: 327–0 × 1–2 × 5–8
Method 1   1–2 × 5–8 = 6–8
           327–0 × 6–8 = 2,180 CF

Method 2   327–0 × 5–8 = 1,853–0
           1,853–0 × 1–2 = 2,162 CF

By decimals: 327 ft × 1.17 ft × 5.67 ft
Method 1   1.17 × 5.67 = 6.64
           327.0 × 6.64 = 2,171 CF

Method 2   327 × 5.67 = 1,854
           1,854 × 1.17 = 2,169 CF

The exact answer is 2,161.84 CF.

There are many legitimate shortcuts, most of them simply a matter of seeing combinations of figures that lend themselves to speedy computations. A little thought is also helpful. In every take-off, there are dozens of opportunities to combine items to simplify the extensions. Remember, the number of steps not only saves time, but also reduces the probability of error.

The take-off for concrete beams might read:

Concrete beams

			Conc	Forms	
L	W	D		Bottoms	Sides
122–0 ×	1– 4 ×	2–2	594 CF	163 SF	1,102 SF
132–6 ×	0–10 ×	2–2		111	
114–8 ×	0–10 ×	1–1½	138	96	272
27–0 ×	1– 0 ×	1–1½		27	
			732 CF	397 SF	1,374 SF
		=	27 CY		

The method used in the above example eliminated eight calculations, reducing the number of steps from sixteen to eight. The extensions on the left are for the concrete, those on the right are for the formwork for beam bottoms and beam sides, respectively. First the formwork items for beam bottoms (163 SF and so on) are extended. Each of these items is also the first step in computing the concrete (L × W). Using these items, therefore, go on to compute the concrete items. The first two items have a common beam depth of 2–2, so the 163-SF area may be added to the 111 SF, giving 274 SF times 2–2, or 594 CF. Repeat this procedure for the other two items: 96 SF plus 27 SF gives 123

SF; 123 SF times 1 ft 1½ in. gives 138 CF. Finally, the beamside forms are computed by first combining those having a common depth; thus the first two beams are handled in one item—254–6 times 2–2, doubled (for two sides). It should be noted that fractional parts of a foot are adjusted to the nearest whole number only in the final figures. Also, if computing only two out of three dimensions—as for beam bottoms—a pencil should be laid over the column of figures not being used, to prevent "jumping" to the wrong dimension.

It should further be noted that beams of the same width and depth were combined into one item before being entered on the take-off sheet. This is yet another time-saving device; the value of a preliminary "collection sheet" will be shown in the take-off sections that follow.

The general idea is to "use your head," making every calculation work for you, to save time and increase accuracy. Quantities to be used for estimating do not have to be set out in the same way as quantities for purchasing schedules. This book is primarily intended to show how to take off items for estimating; it is allowable to combine quantities that will carry a common unit price. A great deal of time is wasted by taking off bidding quantities in dozens of separate items that will be priced at the same units. When preparing the project schedules, it may be necessary to take off, say, glazed brick or structural glazed tile by rooms; but for estimating, such a room-by-room breakdown would be a waste of time. Only items that will require special consideration when the estimate is being priced should be taken off separately. It should also be noted that there are numerous electronic and mechanical "gadgets" that can make the take-off process easier and faster. Some of these are the electronic counter that counts by touching the drawing sheets for typical details, a scaled measuring wheel that can be used to measure lengths by rolling on the drawing sheets, and an electronic calculator that will convert from inches to decimals and vice versa. Since these gadgets are always being developed and are not the thrust of this text, they will not be discussed in detail. I suggest that you consult the trade publications appropriate to your estimating work for details.

Collection sheets

The purpose of these sheets is to collect together items that are similar, to simplify the take-off sheets, save time, and reduce the probability of error. Collection sheets (Table 1.2) are not scrap paper to be thrown away after the items have been transferred to the take-off sheets; they are part of the take-off, and should be retained in the take-off folder. Consider, for example, concrete foundation walls for a medium-sized building—say, 20,000 SF of floor area. There may be variations in thickness and height of exterior walls; wall footings probably would vary; there might be a brick shelf that varies. The exterior concrete walls, footings, forms, and brick shelf could be taken off in hundreds of separate pieces, with each item being laboriously computed. But the collection sheet will do much of that work for you. It will bring together all the concrete foundation walls 12 in. thick, all the 15-in. walls, and so on; moreover, it will bring together the 12-in. walls 4 ft high, the 12-in. walls 5 ft 3 in. high, and so on. It will enable the wall footing items to follow the wall

General Methods and Procedures 15

TABLE 1.2 Collection Sheet for Foundation Walls

Foundation walls

	14 in.			16 in.	
3–9	4–2	5–8	7–2	4–2	5–4
121–4	21– 8	84–8	18–7	19–0	31–6
13–2	17– 9	19–2	43–5		
19–7	32–11	13–6	62–0		
35–4	15– 6	11–4			
189–5	87–10	128–8			

take-off; it will collect the various pilasters; it will reduce the brick shelf to a few items of specific heights. Ordinary 8½-in. × 11-in. white ruled paper, turned sideways, is excellent for "collecting." For very large jobs, any large ruled or squared paper may be used.

After the various quantities that make up an item have been collected, the total is transferred to the take-off sheet. For the concrete foundation walls described in the previous paragraph, the collection sheet would read as shown above.

If there were openings in the walls, they would be entered on the collection sheet and transferred as deductions to the take-off sheet. From the figures shown above, the take-off sheet would read:

Ext. fdtn. walls

```
14 in.   189– 5 × 3–9 =  710 SF  ⎫                           FORMS
          87–10 × 4–2 =  366    ⎬  2,249 SF = 2,624 CF      2 × 2,496
         128– 8 × 5–8 =  729    ⎪                            = 4,992 SF
          62– 0 × 7–2 =  444    ⎭

16 in.    19–0 × 4–2 =   79     ⎫    247   =   329
          31–6 × 5–4 =  168     ⎬
Perim.   518–5                                    
                                    2,496 SF = 2,953 CF
                                             = 109½ CY
```

Note that sixteen figures are collected into six items. Also note that on the take-off sheet the wall surface areas are extended and totaled before being multiplied by the wall thickness. The total perimeter of the walls is computed (518–5) and so becomes available for several items that will come up later. The wall surface area is doubled to obtain the contact area for formwork.

The extensions made in the above example are worth noting:

1. 189–5 × 3–9. Without rewriting the figures,
    189–5 is multiplied by 3   = 568–3
    Since 0–9 is a quarter of 3–0, add   ¼ × 568–3 = 142–1
                                                         710–4 = 710 SF

2. In the other items, inches were also converted to fractions of a foot; for example, 87–10 × 4–2:

$$87\text{–}10 \times 4 = 351\text{–}4$$
$$+\ 87\text{–}10 \times 1/6 = 14\text{–}8$$
$$366\text{–}0 = 366\ \text{SF}$$

SUMMARY OF GENERAL RULES FOR TAKING OFF

Always start in the same place on each drawing and progress around the building in a particular direction. (The examples in this book generally start the take-off at the bottom left-hand corner of a drawing and proceed clockwise.)

Do not be afraid to mark up the drawings as you clear a piece of wall or an item. Check off everything that you have completed, using a small check mark. Check off details, sections, completed drawings (that is, drawings from which you have taken off everything), items on the specifications take-off, items on the collection sheets that have been transferred to the take-off, and items on the take-off sheets transferred to the estimate. Check off each item immediately after the entry has been made.

Always take advantage of duplications of design. Is the building symmetrical about its center line? Are there two or more structural floors with the same design? Do certain beams repeat through two or more floors? Do the room layouts repeat either floor to floor or in different sections of a floor? What features do the rooms have in common?

There is design repetition somewhere in most buildings. It might be nothing more than similar bathrooms in the north and south wings, or it might be the whole floor layout of a hospital that is exactly the same on each of several floors. If there are only slight variations from floor to floor, then the items that vary should be taken off before computing the repetitive items. It might be helpful to hold similar drawings up to the light to compare. Make an item work for you as many times as it is repeated. If you see that the interior partitions in the north and south wings are alike and that they are duplicated on the next two floors, then you can handle two wings on each of the three floors—that is, six wings—all alike. Mark one of the wings on the first applicable plan "6 times" or "1 × 6" in bold colored pencil, draw cutoff marks through the plan to delineate the repeated section, and check off all the five duplicate areas on the three drawings. Next, take off the partitions in the one wing [on a collection sheet, Table 1.3)], add up the completed items, and multiply each total by 6; the six wings are thus all taken off. Probably the best way to mark off the five wings that will not be used is to cross out those areas on the drawings with a large X, as in Fig. 1.1.

The "1 × 6" notation at the right is checked off as the items are done. The first check mark is made when one wing has been completed, and the other when the items have been multiplied by 6.

Great care must be taken in using this "times-ing" method, but properly applied and carried out it is of tremendous help in both saving time and reducing the probability of error. The notation "6 times" or "1 × 6" must be

TABLE 1.3   Collection Sheet for Partitions

Concrete-block partitions

	4-in. ptns.		8-in. ptns.	
	9–2	10–6	9–2	10–6
	31– 0	18– 2	111–0	82–4
	42– 6	16–10	27–0	92–8
	93– 4	24– 0	58–0	32–0
	166–10	59– 0	196–0	207–0
6 wings	× 6	× 6	× 6	× 6
✓	1,001– 0	354– 0	1,176–0	1,242–0

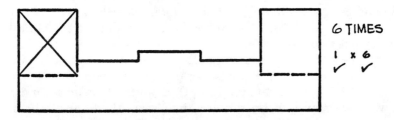

Figure 1.1   Duplication of floor plan.

large enough to stand out, and the check marks must be made as soon as the clearance is completed. The collection sheet must be neat and must clearly show the multiplication by 6, as in the example shown.

Should further repeating items occur, list them under the applicable items, bracket them for totaling, and then add on the "similars" total (that is, the total of this set of items times the number of duplicates). If, for example, this set of items applies to four areas, you have one area already listed and only have to add three times the total of this set of items (with a notation "Add 3"). You cannot multiply beyond your first group, except by carrying the subtotals out into another column:

	166–10	
6 wings	× 6	
	1,001–0	
	22–6	
	14–4	
	18–3 }	147–9
	92–8	
Add 3	443–3	
	1,592–0	

or

```
166–10  × 6  =  1,001–0
 22–6
 14–4
 18–3
 92–8
─────
147–9  × 4  =    591–0
              ─────────
                1,592–0
```

Look for the predominant item of design or detail and make it work for you. For example, the structural floors may be 7-in. concrete slabs for the most part, with small areas of 4-in., 5-in., and 6-in. slabs. Take off the total slab areas—floor by floor—and add them up. You will need the gross floor area anyway for several other items. Then take off all the minor areas of 4-in., 5-in., and 6-in. slabs and deduct their total from the gross slab area to obtain the area of 7-in. slabs. This method is quicker and better than taking off each separate area of 7-in. slab.

Practice adding two items simultaneously, as, for example, in adding windows taken from the elevations. Let us say that there are six types of window in the building, designated A, B, C, D, E, and F. Instead of counting each type of window separately, try two at a time; carry the totals for both types A and B simultaneously. Starting at the top left-hand corner of the elevation sheet, you come first to an A (that is, "1 and 0"), then a B (that is, "1 and 1"), two A's ("3 and 1"), a B ("3 and 2"), and so on. As you count, check off each window with a colored pencil. After a little practice it will become quite simple, and you can step it up to doing three types at a time. It is just a little trick of concentration and practice. (Four items at a time is rather difficult—and is not recommended.)

Do you know the "27 times" table yet?

Always write your items in a set order. Throughout this book, all items are written in the order of length times width times height (or depth); any combination of two items also follows that order.

Generally, an item should indicate all three dimensions somewhere; if not in the quantity, then in the description—or between the two. For example:

Concrete in ext. walls = 27 CY

1-in. T. & G. roof boarding = 1,200 SF

2-in. × 4-in. plates = 1,840 LF

To summarize the basic rules:

Learn the "27 times" table.

Use a red pencil for deductions and queries.

Underline in blue pencil the final total of every item on the take-off sheets—that is, the items that will go onto the estimate sheet.

Make a specifications take-off.

Always check off an item or a drawing as you clear it.

Measure everything as it shows.

Take off everything that is shown.

If it is different—keep it separate.

Make the collection sheet work for you.

Chapter

2

Excavation and Site Work

Excavation is a difficult trade to take off. By contrast, structures are of definite and fixed dimensions; the take-off for them follows a clearly defined procedure and method. Taking off excavation, however, entails several unknown factors which require knowledge and judgment. Consider, for example, that a hole is to be dug; the calculation is simple. But how big is the hole to be? Will the banks stand up? What slope should be figured? What may be reasonably inferred from the borings? Where is the water level? How is the drainage? Is there topsoil to be stripped? If so, how deep is it? For what is the excavated material to be used? How much of it can be used for fill? Is it going to be trucked away? If so, where to? Is the surplus excavated material enough for the fill areas? Does the material to be excavated meet the requirements for backfill, for general fill, or for subgrades for pavings? These are only a few of the questions that must be answered in taking off the excavation.

Excavating never conforms to exact quantities; in fact, there is always an element of "intelligent approximation" involved. There can be no certainty about the actual width for a trench, and even if there were, no machine or shovel operator would be able to dig it dead to line. Nor can one be certain about the extent to which excavated material will bulk up when in the loose. Ten CY in the ground might be 12 CY on the truck, or perhaps 11 CY, or—who knows? Knowledge of the local area, study of the borings, previous experience, and examination of the site will all help in estimating the increase in bulk—but that is what it is: an estimation, an approximation.

The excavation take-off should not be started until all the substructure concrete has been taken off. Whether the excavation is taken off immediately following the below-ground concrete, after the entire concrete work, or even after all trades for the entire building proper, is a matter of choice. But you cannot determine the problems of the building excavation until you have taken off the foundations, the ground slabs, and all the rest of the substructure concrete.

VISITING THE SITE

Always try to inspect the site before starting the excavation take-off. Familiarity with the site will often help you to decide how to handle some of the excavation problems. Often it is possible to visit a site when picking up the bid drawings or while en route to another job or appointment, but sometimes a special trip will be necessary. It might also be necessary to revisit the site when the estimate is ready to price. The problems encountered and the overall value of the excavation and site work should determine whether two visits to the site are justified.

For many reasons, it is imperative that the site be visited. The plot plan should show all existing site conditions, but plot plans are sometimes so sketchy that one must examine the site for anything—old foundations, curbs, buildings, trees, shrubs—that will have to be removed. It is quite common to discover that a site is completely wooded, although no trees have been shown on the plot plan.

Determining the depth and quality of the topsoil may present a problem. For bidding jobs that include considerable excavation or a fair amount of topsoil stripping, a hand shovel should be taken along when visiting the site. The entire bid may turn on one question: how much topsoil is there and is it good enough to meet the requirements for lawns? The quantity of topsoil affects other aspects of the job too. On a site-cut job (for which the grading is essentially cut) more topsoil means less cutting. If fill must be added to raise a low site, however, then the more topsoil to be stripped the more fill to be supplied.

Rock is another vital consideration. If rock excavation is specified as an extra-cost item over the amount of the contract, the unit price for rock may be very important. A given unit price (that is, a unit price embodied in the specifications by the architect) that is very low may have to be compensated for in the bid if there is much rock on the site. Or a large amount of rock excavation with a good unit-price extra may cause a contractor to tighten his or her bid. The size of boulders not classified as extra-cost rock may also be important. For example, a certain architect or engineer may include boulders up to 4 CY with the unclassified "earth" excavation. The amount of rock depends on the nature of the site, but on a recent bid for which the specifications included this provision, rock outcroppings were found all over the site. A 4-CY boulder is quite an item—8 tons of rock at no extra cost!

Of course, the general nature of the ground will have to be considered. A good set of borings will help you; two or three test pits may only fool you! If other construction is going on near the site, you may be able to see how the banks stand up, what the water condition is like, and what the ground material is like. All these observations can help, but cannot be conclusive.

Access to the site must also be evaluated. Are you going to be hemmed in? Will you have to construct a temporary access road? Will you be tight for storage room? How far is the nearest electric power line? How far to the water main? Where is the telephone line?

Utilities may present several problems. Where do your lines go to from the building? Where do the permanent power, water, and gas supply lines, fire alarm, and telephone lines come from? Will you have street digging, or road and sidewalks to cut and make good? Will there be any utility companies to pay? Any city fees for street work? Any traffic protection or police control to be paid by the contractor when working in the public road?

Surplus earth may have to be removed from the site. Where is it to go? Does it contain hazardous material? Can it be sold? Is there a dump handy? Is there a charge for dumping? How many miles is it to the dump?

Fill may have to be transported to the site. Is there a pit nearby? If so, what is the material like? How will it compact?

These are only the most common problems that must be considered. Particular jobs may have their own site problems—anything from radio control for hauling earth around airports to problems of logistics for a job in a remote area.

Site clearing

If possible, all items to be removed should be measured; otherwise they should be described in detail so a lump-sum price can be developed. Walls, curbs, fences, and the like can be taken off in lineal feet; isolated trees can be counted. Wooded areas are measured in acres, with some indication given as to the extent and nature of the trees, such as "heavily wooded" or "small trees" or "mostly scrub."

The site area will probably be taken off at this time. The limits of contract lines on the plot plans should be clearly marked with colored pencil. If the site is irregular in shape, it may have to be broken up into triangles to compute the area—a procedure called "triangulation." An irregular site with curved boundaries might require offsets for calculating the area.

To compute the area of a site that has been split into triangles, note the areas that have a common base. In Fig. 2.1, triangles b-c-d and b-e-d have a common base (b-d). Those two triangles may be computed as one item: the length of b-d times the average perpendicular height. Scale off the perpendicular height of each of the two triangles, add them together, and divide by two for the average height, which multiplied by b-d will give the total area of the two triangles.

Short irregular boundaries may be treated as straight-line boundaries in marking off the limit lines, as in Fig. 2.2. The curved lines are the actual site boundaries; the dotted lines are the boundaries superimposed for computing the area.

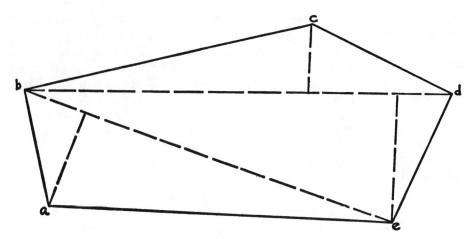

Figure 2.1 Determination of site area. Triangulation.

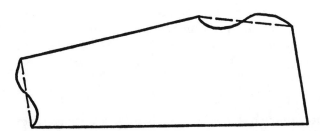

Figure 2.2 Determination of site area. Irregular boundaries.

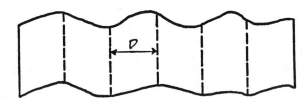

Figure 2.3 Determination of site area. Curved boundaries.

For irregular areas having more or less parallel curved boundaries, as in Fig. 2.3, the area of the site may be expressed as the length times the average of the seven offsets. Or, for greater accuracy, the area could be computed by the midordinate method:

D = Distance between offsets
S = Sum of intermediate offsets

$$\text{Area} = D\left[\frac{(\text{1st offset} + \text{last offset})}{2} + S\right]$$

Still another method is "Simpson's Rule of One-Third," which is more complicated and is not usually used for estimating.

Stripping topsoil

As in all excavation work, there is an element of guesswork in this item; not blind guesswork, but without certainty. The depth of the topsoil shown on the borings can be averaged, the areas in which topsoil is shown can be marked off, the site can be examined, and certain conclusions may thus be drawn. But a great deal depends on the quality of the topsoil and the nature of the site. What is classified as topsoil on the borings may prove to be very poor topsoil sand when you start stripping. If the site is wooded with fair-sized stumps to be grubbed up, much topsoil may be lost in loading roots and stumps to haul off the site.

The quantity of topsoil required for the lawns and planting may affect the take-off for topsoil stripping. What happens to the surplus topsoil: does it become the property of the owner or the contractor?

Topsoil stripping is taken off in cubic yards, with the thickness stated in the description: "Strip and stockpile topsoil (8 in.)—...CY."

BUILDING EXCAVATION: EXAMPLE 1

The building excavation must be taken off in several separate items, such as: "Machine bulk excavation, building; Machine trench excavation, building; Hand excavation, footings; Excavate pier footings." The bulk excavation is measured to the underside of the basement slab or to the underside of the gravel bed for the slab, allowing for the depth of topsoil stripping already taken off. The top grade must be averaged from the grades shown on the plot plan. The allowance for working space depends on the depth of the excavation and the nature of the material. The excavation can seldom be less than 3 ft outside the building lines, but the distance cannot be determined exactly. It is often useful to set up the excavation lines on a wall section. This setting up takes only a few minutes, and then, knowing the type of material (from the borings), you can see what appears to be a reasonable slope for the banks. If the excavation cut is to be sheeted, of course there is no problem of banks or slope.

An excavation line superimposed on a foundation-wall section is shown in Fig. 2.4. The average width of the bulk excavation is 3 ft 9 in. outside the building line; hand work for the wall footing is taken off separately.

Working Drawing 2.1 shows the foundation walls of a building, with the wall section the same as that in Fig. 2.4. The perimeter and the floor area are given because they would have been taken off and determined as part of the foundation concrete.

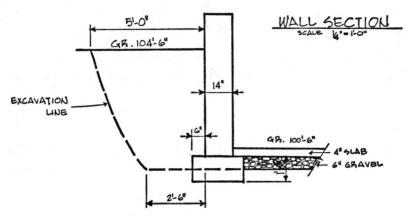

Figure 2.4 Section through foundation wall.

The take-off (W. D. 2.1)

Machine bulk exc. bldg.

$$
\begin{aligned}
&\qquad\qquad\qquad 4{,}534\ \text{SF} \times 4\text{–}10 = 21{,}912\ \text{CF} \\
&\text{Perim.}\quad 309\text{–}0 \times 3\text{–}9 \times 4\text{–}10 = \underline{\ 5{,}602\ } \\
&\qquad\qquad\qquad\qquad\qquad\qquad\quad\ 27{,}514\ \text{CF} \\
&\qquad\qquad\qquad\qquad\qquad\qquad = \underline{\underline{1{,}019\ \text{CY}}}
\end{aligned}
$$

Hand exc. wall ftgs.

$$
\begin{aligned}
289\text{–}4 \times 3\text{–}0 \times 0\text{–}6 &= \underline{434\ \text{CF}} \\
&= \underline{\underline{16\ \text{CY}}}
\end{aligned}
$$

Backfill: walls & ftgs.

$$
\begin{aligned}
&\qquad\qquad\ \text{Bulk perim.} \qquad\quad = 5{,}602\ \text{CF} \\
&\text{Ftgs.}\quad 289\text{–}4 \times 0\text{–}10 \times 0\text{–}6 = \underline{\ \ \ 121\ \ } \\
&\qquad\qquad\qquad\qquad\qquad\qquad\ \ 5{,}723\ \text{CF} \\
&\qquad\qquad\qquad\qquad\qquad\ = \underline{\underline{212\ \text{CY}}}
\end{aligned}
$$

Hand trim for ground slab   = $\underline{4{,}196\ \text{SF}}$

Surplus material to remove

$$
\begin{aligned}
&\qquad\qquad\qquad\qquad\qquad\qquad 823\ \text{CY} \\
&+\ \text{Bulkage 15\%}\quad\ \underline{\ 123\ } \\
&\qquad\qquad\qquad\qquad\qquad\ \underline{\underline{946\ \text{CY}}}
\end{aligned}
$$

Notes on the take-off (W. D. 2.1)

The perimeter is separated from the building area in order to have the quantity for backfilling available without having to recalculate it. The method by which the perimeter and floor area are obtained is explained in Chap. 3. The gross building area is the floor area plus the perimeter wall area—4,196 SF—plus (289–4 times 1–2); that is, 4,196 plus 338, or 4,534 SF. The depth of 4–10 is from the present grade to the underside of the gravel bed.

The length of the perimeter item is figured as follows:

	Wall perimeter				=	289–4
+	to outside lines	4	×	1–2	=	4–8
+	to excavation line	4	×	3–9	=	15–0
						309–0

The hand-excavation length is the same as the wall perimeter, since the footing excavation is equidistant from both sides of the wall. The footing is 2–2 wide, so an excavation 3 ft wide would be sufficient to allow for erecting the edge forms.

The take-off for the backfill is the perimeter quantity (5,602 CF) plus the backfill around the footing. The footing backfill is the 10-in. working space included in the 3-ft-wide footing excavation. Very often, having taken off the concrete, we can use those quantities to figure the backfill. In this instance, reference to the concrete quantities would show the wall footings (289–4 × 2–2 × 1–0) to be 627 CF. The hand excavation item is to half of the 12-in.-deep footing, so the concrete used in that space will be 313 CF. Deducting that figure from a total excavation of 434 CF gives 121 CF to be backfilled.

The item for hand trim is the floor area that must be graded before pouring the slab. The actual leveling may be to the ground or simply of the gravel fill, but whichever it is, an item has to be carried for it.

The surplus for removal is the total excavated material less the quantity needed for backfill. An allowance must be made for the increase in bulk when the material is in the loose. For rock on trucks, the increase may be as much as 30 percent. Normally the balancing of excavation and backfill is not made until all excavation work has been taken off. It is shown here as if there was no further excavation, but on most jobs there would be site cut and fill still to take off, plus various site-work excavations, before striking the balance.

BUILDING EXCAVATION: EXAMPLE 2

The foundation plan and section are shown in W. D. 3.1 for a building whose slab-on-ground is above the natural grade. Since the excavation take-off will follow the quantities for the foundation items, the concrete take-off (p. 59) should be read in conjunction with this example.

28 Chapter Two

The excavation will be for wall trenches (no bulk excavation) and footings. The footing is 2 ft wide, and so the trench should be 4 ft wide at the bottom. At a slope of about 1 in 2, the trench will be 8 ft wide at the top, for a 6-ft average width. The wall bottom varies. The trench excavation item is taken off to within 3 in. of the bottom of the footing, and that last 3 in. taken off as hand work.

The take-off (W. D. 3.1)

Strip & stockpile loam *
$$107\text{-}0 \times 77\text{-}0 = 8{,}239 \text{ SF} \times 0\text{-}6 = 4{,}120 \text{ CF}$$
$$= \underline{153 \text{ CY}}$$

Machine trench exc. bldg.
$$314\text{-}10 \times 6\text{-}0 \times 5\text{-}0 = 9{,}445 \text{ CF}$$
$$= \underline{350 \text{ CY}}$$

Wall area	= 2,232 SF
Wall perim.	= 314-10
Average height	= 7- 1½
101- 3	- 2-10½
- 98- 4½	4-3 below grade
2-10½	Add 0-9 footing
Exc. height	= 5-0

Exc. pier ftg. †
$$12/\ 8\text{-}0 \times 7\text{-}0 \times 5\text{-}0 = 3{,}360 \text{ CF} = \underline{125 \text{ CY}}$$

Hand exc. wall ftgs.
$$314\text{-}10 \times 3\text{-}0 \times 0\text{-}3 = \underline{9 \text{ CY}}$$

Backfill walls

Total exc. material = 484 CY

Wall 314-10 × 1-0 × 4-3 =	49½ CY
Wall ftg.	23½ 83
Pier ftg.	10
	= 401 CY

Fill to ground slab
$$4{,}777 \text{ SF} \times 1\text{-}6\tfrac{1}{2} = 7{,}365 \text{ CF} = \underline{273 \text{ CY}} \text{ (see below)}$$

Site material	= 83 CY
Material to buy	= 190 CY
+ 15%	29
	219 CY

* Taken off approx. 8 ft outside max. bldg. lines.
† The oblique line (/) is a multiplication sign (an "×" may be used if preferred). The only advantage in using the "/" sign is to differentiate the quantity from the dimensions, so that one can quickly see, for example, that there are 12 footings — each 8-0 × 7-0 × 5-0.

Notes on the take-off (W. D. 3.1)

The topsoil strip is taken off to a rectangle approximately 8 ft beyond the building lines. The extent of this item would be indicated on the drawings or in the specifications. The excavation for wall trenches is based on a width of 8 ft at the top and 4 ft at the bottom, an average width of 6 ft. For the depth of the trench, note that on the concrete foundation wall take-off (p. 59) there is 314–10 of wall with a total surface area of 2,232 SF. Dividing 2,232 SF by the 314–10 perimeter gives an average depth of 7–1½. The distance from the top of the wall to the grade after the topsoil stripping is 2–10½, leaving 4–3 of wall below ground. Adding 9 in. for the footing gives a total depth of 5 ft for the trench; this leaves 3 in. of hand excavation.

The pier footings item follows the grades of the walls nearest to them (as can be seen in W. D. 3.1); therefore they too must be 5 ft below grade. The average size of the 12 pier footings is 4×3 ft (or 12 SF of "on the ground" area). The 2-ft wall footing required a 6-ft trench, so a 4-ft footing for the same slope would require an 8-ft trench, and the 4×3-ft footing an 8×7-ft hole. Thus 12 holes are required, each $8 \times 7 \times 5$ ft. The take-off for backfill starts with the total excavation (350 plus 9 plus 125) in cubic yards, from which the concrete is deducted. The concrete wall is 4 ft 3 in. in the ground, so there is 49½ CY in that item. The wall footings are entirely below ground. The backfill item for piers and pier footings consists of all the footings (8 CY) plus 2 CY of the 3 allowed for the piers.

The fill to ground slab item is the floor area times the distance from the underside of the gravel bed down to the natural grade after topsoil strip. This gives 273 CY, but since some surplus earth is available at the site, that quantity is broken down as follows: 83 CY of site material, and 190 CY to buy. To the 190 CY to buy is added 15 percent for truck measure (which is the way the material is bought); thus, 190 plus 29 CY is 219 CY to buy.

ROCK EXCAVATION

Rock excavation must be measured as accurately as is possible from the available information. The excavation lines for rock do not require quite so much working clearance as earth excavation. In general, the rock excavation line at the bottom would be the same as for earth, but the banks would not be sloped. If the wall section in Fig. 2.4 was to be in rock, for instance, the excavation lines would extend 2½ ft outside the wall lines. Rock cannot be excavated to an exact grade, and any attempt to do so would almost inevitably result in failure to "get the bottom" in many places and incur almost a double expense for redrilling. In order to be reasonably sure of getting down to grade everywhere, the rock should be drilled to an extra depth of 6 in. or more, depending on the nature of the rock and the depth being drilled.

For the building shown in W. D. 2.1 and the wall section in Fig. 2.4, the rock excavation would be taken off to 6 in. below the gravel (that is, to the bottom of the wall footing) and around the perimeter the rock would be drilled 6 in. deeper (say, 4 ft wide) to ensure getting the footing.

30   Chapter Two

The take-off (W. D. 2.1)

Exc. rock (bulk) — bldg.

	Bldg.				=	4,534 SF		
+ Perim.	304–0	×	2–6		=	760		
						5,294 SF	× 5–4 =	28,235 CF
+ Ftg.	289–4	×	4–0	× 0–6			=	579
								28,814 CF
							=	1,067 CY

Load & truck rock from site

             1,067 CY
     + Bulkage 40% 427
             1,494 CY

Level off rock for footing

     289–4 × 2–6 = 724 SF

Gravel bed for ground slabs

     4,196 SF × 12 in. = 156 CY
      + Compaction 15% 24
             180 CY

Backfill — walls & ftgs.

   Perim.   760 SF × 5–4 = 4,053 CF
  + Ftg. (all)   579
             4,632 CF
           = 172 CY

Notes on the take-off (W. D. 2.1)

  The excavation area is the building area previously determined (p. 27): the floor area (4,196 SF) plus the walls (289–4 × 1–2)—4,534 SF, plus the working-space perimeter. That outer perimeter is:

	Wall perimeter				=	289–4
+	to outside wall lines	4	×	1–2	}	14–8
+	to excavation lines	4	×	2–6		
						304–0

The depth is from grade 104–6 to 99–2 (5–4), which allows for excavation 6 in. below the bottom of the gravel bed.

The trucking allows an increase of 40 percent in the rock volume for bulking up. The bulk-up allowance will vary considerably; much depends on how the rock is broken up. The larger the pieces, the lesser the yardage per truckload. One or two large, irregularly shaped pieces of rock can make a truckload—a load consisting half of voids, but nevertheless a load.

Some leveling for the footings will usually be needed. If the walls are to rest on rock without footings, it may be necessary to take off items both for leveling and for doweling the walls to the rock.

The gravel bed is taken off at 12 in. deep because the rock excavation item was taken off an extra 6 in. deep. Actually, the amount of gravel required will vary according to how the rock fractures when dynamited. The projection of the footing under the slab is not deducted from the gravel quantity.

The backfill is taken off only for the outside of the walls. The 12 in. taken off for the gravel bed will allow sufficient surplus material to fill the small area around the inside of the excavation at the footings.

Like most excavation items, rock excavation involves many unknown factors, and the quantities are at best a close approximation. Hard granite with the seam diagonal to the drilling line may break, leaving a toe, so that after a few unlucky blasts it becomes necessary to drill deeper to remove the toe. It is usually considerably cheaper to drill a little deeper the first time.

Even if rock excavation is specified at extra cost over contract, it might be necessary to take off rock quantities. In figuring a unit price for rock, you should check the specifications for payment lines. Pay lines that are too tight must be allowed for in your own unit price for extra-cost rock; if the unit price has been set up by the architect, pay lines are just as important to your thinking. You may be paid for rock excavation as measured to a pay line 2–0 outside the building lines, but in actuality have to excavate 3–0 outside the building lines to allow for footings and formwork.

Rock excavation in trenches must be taken off separately from open rock excavation. Rock excavation where blasting is prohibited must also be taken off separately. Similarly, separate any rock excavation that must be blown in small pieces, such as when blasting close to another building.

Sheeting for excavations

Sheeting for excavations is measured in square feet, allowing sufficient length both for driving the sheeting into the ground below the excavation grade and for the top of the sheeting above ground. Driven sheeting will "go" according to the material and obstructions encountered, and so the sheeting will be up and down; thus the length figured should never be less than the next even-foot length (that is, 6 ft, 8 ft, 10 ft, and so on) greater than the net depth of the excavation.

Special sheeting, such as steel columns, solider beams and lagging, interlocking steel sheeting, or something similar, should be taken off in detail and fully described, as in the following: "H-cols. for sheeting (50 pcs. at 18 ft)—

31.7 tons; Interlocking sheet piling (2,200 SF)—33.4 tons; I-beam spreaders for trench sheeting—4.1 tons."

Ordinary plank sheeting of the type used in a sewer trench of moderate depth and width, is measured as the net square-foot face area of the trench, doubled for the two sides, and the item notated "b.s.m." (both sides measured).

SITE GRADING

Cut and fill for site grading is an item for which the quantities will vary considerably; even if taken off twice by the same person, the totals might not be exactly the same. On a scale of 1 in. to 40 ft, each 1-in. × 1-in. square represents 1,600 SF, so that in estimating the cut or fill for that area, there is a variation of 5 CY of material for every 1-in. difference in depth. It is impossible to take off cut and fill to within 1 in.—in fact, those items are usually measured in multiples of 3 in. or 6 in. It should also be noted that the finished area when graded will probably vary slightly from the drawings. If road profiles are shown on the drawings, the cut and fill for roads should be taken off separately and measured more accurately than general grading.

General grading is usually taken off from the site plan by blocking off the drawing into suitable squares and taking each square separately. But before considering the actual amount of cut and fill, there are other factors that must be determined. The topsoil stripping would already have been taken off, and now the depth of the topsoil to be stripped and the specified depth of the topsoil for finish grading must be compared (Fig. 2.5). If the topsoil stripped was the same depth as the topsoil required for finish grading, there would be no problem, because the difference between the present grades and the finish grades would be the same as the difference between the present grade after stripping and the finish subgrade. If there was no topsoil to strip, however, and 6 in. of topsoil was to be allowed for finish grading, then cuts would be increased by 6 in. and the fill decreased by the same amount. If, on the other hand, there was 12 in. of topsoil to strip but only 6 in. required for

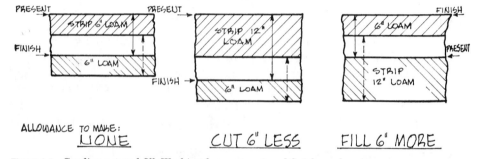

Figure 2.5 Grading cut and fill. Working from present and finish grades.

finish grading, then cuts would be decreased by 6 in. and fill increased by that amount.

One other factor should be considered before taking off cut and fill. Any peat, silt, or other spongy or unstable material specified for removal must be taken off and notated "remove from site" if it has to be hauled away. The extent of that item will have to be allowed for in taking off the cut and fill.

In taking off cut and fill by gridding the site, it is better to use a transparent plastic sheet than to mark off the drawing itself. Heavy plastic sheets can be bought marked off in 1-in. squares, which, laid on the drawing, will save time and prevent the drawing from becoming so marked up that it cannot be read. It is also helpful to mark off the plastic sheet with letters for one direction and numbers for the other, so that each block is identified.

Working Drawing 2.2 shows an area to be graded ("Topsoil stripped 6 in.; allow 6 in. for topsoil for finish grading"). The broken lines are present grades, and the solid lines finish grades. The drawing is marked off in 1-in. squares, each square (or block) representing 40 ft × 40 ft, or 1,600 SF. The letters along the top and the numbers down the side identify the squares: A1, B1, C1, and so on. Starting at A1 and working across the drawing, we would consider each square separately. We might take off the present grade at each corner of a block to obtain an average present grade, and then do the same for the finish grades, with the difference between the two averages being the cut or fill needed for that square. This is a laborious and slow method, however, and, despite all the work involved, the result is still an approximation. It is much simpler to take the present and finish grades at the center of each block, allowing for any abrupt changes, and so determine each cut or fill depth in a single mental calculation.

Let us compare these two methods, taking at random three blocks—say A1, B3, and B5 (see tabulation below).

The three blocks taken off at the four corners (4.55 ft plus 4.35 ft plus 2.73 ft) show a total cut of 689 CY, while the three blocks taken the simpler way (4.5 ft plus 4.5 ft plus 2.5 ft) give a total cut of 682 CY; the difference of 7 CY is negligible for grading quantities. It must be remembered that grading cut-and-fill items are always approximate quantities. The present grades shown are themselves approximations, being simply contours drawn in to connect spot grades taken at intervals. The finish grading will be done to grade stakes, and will not necessarily conform to the exact contours shown on the drawing; between the stakes, the grades will be shaped to pleasing contours. Thus the simplest and fastest satisfactory way in which to take off cut and fill is by using the grades at the center of each block. The collection sheet for W. D. 2.2 is shown on page 34.

Notice that D6 is an example of a block that must be adjusted to allow for abrupt changes in contours. Although there is 4 ft of fill at the center, this decreases to 2 ft at the outside, and so this block is taken off for 3 ft of fill.

For a small site such as that shown in W. D. 2.2, the advantage of using a collection sheet is slight, but the method would be the same no matter how large the site; thus for large grading jobs there would simply be more items in each of the groups, and cut and fill for the entire site could still be taken off in only twenty or thirty calculations. Table 2.1 is the collection sheet for cut and fill.

TABLE 2.1 Collection Sheet for Cut and Fill

Grading (1,600-SF blocks)

	Cut	Fill
0–6	D2	D1
1–0	C4, C5	
1–6	C1, B6	C6
2–0		D5
2–6	C2, B5, A6	
3–0	C3	D6
3–6	B1, B2, B4	
4–0		
4–6	A1, B3	
5–0	A2	
5–6	A5	
6–0	A4	
6–6	A3	Loam strip 6 in.
		Lawns, 6 in. loam

No change — D3, D4

4-Corners Method   *Center Method*

		Present	Finish		
Block A1	1.	90.1	85.1	Present	88.5
1 2	2.	88.3	85.0	Finish	84.0
	3.	88.9	83.5	Cut	4.5 ft
	4.	87.5	83.0		
3 4		354.8	336.6		
		336.6			
		4)18.2			
	Cut	4.55 ft			

Excavation and Site Work

Block B3	1.	86.0	80.5	Present	83.8
	2.	83.7	79.8	Finish	79.3
	3.	84.3	78.8	Cut	4.5 ft
	4.	81.0	78.6		
		335.0	317.7		
		317.7			
		4)17.3			
	Cut	4.35 ft			

Block B5	1.	82.9	78.1	Present	79.7
	2.	79.0	77.6	Finish	77.2
	3.	80.4	76.9	Cut	2.5 ft
	4.	77.8	76.6		
		320.1	309.2		
		309.2			
		4)10.9			
	Cut	2.73 ft			

The take-off (W. D. 2.2)

Site grading — cut

1/	1600 SF	×	0–6	=	800 CF
2/	1600	×	1–0	}	8,000
2/	1600	×	1–6		
3/	1600	×	2–6	=	12,000
1/	1600	×	3–0	=	4,800
3/	1600	×	3–6	=	16,800
2/	1600	×	4–6	=	14,400
1/	1600	×	5–0	}	36,800
1/	1600	×	5–6		
1/	1600	×	6–0		
1/	1600	×	6–6		
					93,600 CF
				=	3,466 CY

Site grading — fill

1/	1,600 SF	×	0–6	}	11,200 CF
1/	1,600	×	1–6		
1/	1,600	×	2–0		
1/	1,600	×	3–0		
				=	415 CY

Material to remove from site

			3,051 CY
+ Bulking up	15%		449
			3,500 CY

Area to subgrade

240 ft × 160 ft = 38,400 SF

Lawns — spread loam (from site stockpile)

38,400 SF	×	6 in. =	19,200 CF
		=	711 CY
	+ 10%		71
			782 CY

Spread lime (10 lb per 1,000 SF)	=	400 lb
Comm. fertilizer (20 lb per 1,000 SF)	=	800 lb
Grass seeding (5 lb per 1,000 SF)	=	38,400 SF
Maintain lawns until 2nd cutting	=	L.S.

Notes on the take-off (W. D. 2.2)

In taking off a job, the entire excavation section would be completed before balancing the excavation and fill items to determine the quantity of material to be removed or bought. In the calculation of the grading cut and fill, certain items are combined to save multiplications; for example, the second and third items of the grading cut (two at 1–0 and two at 1–6) are combined (as one at 5–0), so that 1,600 SF times 5–0 gives 8,000 CF.

The details of the lawn items would be taken from the specifications. The quantities of lime and fertilizer are rounded off to the next 50 lb above the net quantity.

Miscellaneous landscape items

Lawn sodding is taken off in square feet separately from the seeding quantities. Any special lawn items are measured and taken off item by item; for example, swales are measured as an extra-cost item: "Form grass swales—…SF."

Planting is usually treated as a subitem; if taken off, however, the various types of shrubs are listed separately by sizes.

Tree wells are taken off by number and fully described; for example: "Fieldstone tree wells 6-ft dia. × 1–6 deep—8 ea."

ROADS AND PAVINGS

The cut and fill for pavings may be taken off as part of the general grading cut and fill, making allowance for the paving subgrades where they differ from the subgrade for the lawns, thus doing the entire grading in one operation. Alternatively, roads, parking areas, drives, and the like can be marked off on the drawing, and the excavation for them taken off separately from the general grading. Either method may be used, depending on the site layout and personal preference. If road profiles are shown on the drawings, they should be used for the road cut and fill.

Still another method is to first take off the entire site with the general grading—that is, all the exterior areas (such as lawns, roads, and walks)—measured as at the lawns subgrade; then take off the paved areas for extra depth. Thus, if the subgrade for the lawns is 6 in. low and the road subgrade is 14 in. below the finish grade, having previously taken off the entire site as at 6 in. below the finish grade, we can use the road profiles for the additional 8 in. of cut or fill. Cuts for roads should be taken off wide enough to include the curb, or the width of the gravel or stone bed if it extends beyond the actual road width. Fill for roads must be taken off wide enough to allow for the slope of the banks. If a road is to be built on an embankment that varies in height, it will be necessary to take a number of cross sections in order to determine an average cross section for the embankment fill.

When preparing a collection sheet for the roadwork, you should also include thereon the areas for all pavings, driveways, courtyards, curbs, walks, and similar work connected with the pavings. Guard rails, parking lines, wheel stops, fencing, and the like should also be included. Then, using the collection sheet and the cross sections and details on the drawings, complete the take-off sheet.

Working Drawings 2.3 and 2.4 show a partial site plan: a road, a parking area, bitumen walks, concrete walks, and curbs. We note that the road is 30 ft wide and has a granite curb at one side and a bitumen curb at the other. The subgrade for the road ($14\frac{1}{2}$ in. below finish grade) and the distances between points where cut-and-fill depths change are details that we would mark on the profile when starting our take-off for the paving; they are not part of the profile drawing.

This is an example of a plan that does not show the finish contours; the only information on them is shown on the profiles. Therefore we must use the profiles for our take-off.

Assume that the grading (except for the road) has been taken off previously. In this take-off, we shall include cut and fill for the road, the extra excavation for the curb, all the gravel, surface material for all pavings, the bitumen curb, and the granite curb.

The curbing around curves is measured best by using a measuring wheel or a steel scale tape on edge, following the curve. The scale measurement may be converted to the actual dimension by laying it off on the proper engineer's or architect's scale. Since the two radius pieces at the street are not detailed, they should be taken off as scaled (as 5-ft curve pieces). All dimensions not given are scaled off. The collection sheet is shown in Table 2.2.

TABLE 2.2 Collection Sheet for Walks and Roads

Roads
344–0 × 30–0 = 10,320 SF
80–0 × 25–0 = 2,000
                 12,320 SF

Bit. walks
25 × 10 ft = 250 SF
75 × 6   = 450
               700 SF

Conc. walks   Forms
259–0 × 6–0 = 1,554 SF   518 ft
 72–0 × 8–0 =   576   144
 30–0 × 23–0 =   690   68
                    2,820 SF   730 ft

Bit. curb
     52 ft
   220
   272 ft

Gran. curb
    195 ft
     85
     63 (rad.)
     20
2/25 = 50
     80
   493 ft

Pkg. area — grades

	Present	Sub
	34.0	33.8
	35.2	35.8
	37.5	2.0 cut
	36.5	
	4)143.2	
	35.8	

Gran. curb 5-ft rad. corners = 2 pcs.

The take-off (W. D. 2.3 and 2.4)

ROADS AND PAVINGS

Exc. for roads

           33–0 × 31–6 × 1–0 = 1,040 SF = 1,040 CF
+   10–0 × 31–6 × 1–4 =   315   =   420
½/   53–0 × 31–6 × 1–3 = 1,670   = 1,046
½/   70–0 × 31–6 × 2–3 = 2,205   = 2,480
Pkg. 82–0 × 25–6 × 2–0 = 2,091   = 4,182
                                  7,321 SF   9,168 CF
                                                  =   340 CY

Fill for roads

½/   18–0 × 31–6 × 0–3 =   567 SF =   71 CF
½/ 170–0 × 31–6 × 1–4 = 5,355 SF = 3,571
                                 5,922 SF   3,642 CF
                                              =   135 CY

Shape & roll subbase for road

$$\begin{array}{r} 7{,}321 \text{ SF} \\ -315 \\ \hline 7{,}006 \\ +5{,}922 \\ \hline 12{,}928 \text{ SF} \end{array}$$

Grading for walks — taken off previously

Shape & trim subbase for walks

$$\begin{array}{r} 2{,}820 \text{ SF} \\ 700 \\ \hline 3{,}520 \text{ SF} \end{array}$$

Graded gravel base for pavings

Road	12,928 SF	×	12 in.	=	12,928 CF	
Walks — conc.	2,820	×	6	=	1,410	
Walks — bit.	700	×	8	=	467	
+ Curbs	500–0 × 1–6 × 1–0			=	750	
					15,555 CF	
				=	576 CY	
				+	87	(15%)
					663 CY	

2½-in. bitumen paving — roads	=	1,370 SY
2-in. bitumen paving — walks	=	78 SY
6-in. × 6-in. bit. curb	=	272 LF
Edge forms for conc. walks	=	730 LF

4-in. conc. walks (2,500 psi)

	2,820 SF	=	940 CF	= 35 CY
Broom finish conc. walks		=	2,820 SF	
Reinf. mesh for walks		=	? (none shown)	
Cure conc. walks		=	2,820 SF	

Hand exc. for curbs

$$750 \text{ CF} = \underline{28 \text{ CY}}$$

Granite curbs 6 in. × 18 in.	=	430 LF
Granite curbs rad. pcs.	=	63 LF
Granite curbs 5-ft rad. corner	=	2 pcs.

Notes on the take-off (W. D. 2.3 and 2.4)

Road cut and fill is taken off from the profiles. The first item (33 ft × 31–6 × 1–0) is the 33-ft length on the profile measured below the hump; the hump is the "add" item taken off next (10 ft × 31–6 × 1–4). (The 31–6 width is the 30-ft road plus 6 in. at the side that has a bitumen curb and 1 ft to the outside of the gravel at the side that has a granite curb—all as shown on the road cross section.) Generally, the items are triangular in longitudinal section; therefore the depth is measured at the maximum point, and the item is "one-half times."

The areas of the items are converted to cubic feet and totaled, so that they can be used for the gravel take-off and for the shape-and-roll item.

The parking-area excavation is taken off by averaging the present grade at the four corners and deducting that figure from the finish subgrade, as shown on the collection sheet.

Shape and roll is an item that should always be taken off for the road subbase. The walks would have a similar item, but for hand trim rather than roller work.

In taking off the gravel around the curb, the estimator does not deduct for the curb; the 12-in. gravel roadbed is taken off across the entire width of 31 ft 6 in., and an additional 18 in. × 12 in. of gravel is taken for under the curb. For waste and compaction, 15 percent is added to the gravel item. Also an item for hand excavation must be taken off for the trench at the curb, its quantity being the volume of the extra gravel (750 CF).

Notice that when the bitumen paving items are transferred from the collection sheet to the take-off sheet, the conversion from square feet to square yards can be done mentally and the result entered directly on the take-off sheet. A note is made of the fact that there is no reinforcing shown for the concrete walks. The specifications should be checked later for the queried items.

UTILITIES

Before taking off the utilities, you must decide whether they are going to be installed before or after the site grading; that is, whether the utilities excavation is to be figured from present or finish grades. The depth of the trenches and the nature of the ground must be examined to decide where sheeting will be needed. OSHA regulations govern the requirements for sheeting. Usually, trenches over 6–0 deep will be measured for sheeting even if no local law requires it. The width of the trench is governed by its depth, the nature of the ground, and the size of the pipe. City regulations must also be checked for such requirements as fees for connections to mains and permits for street openings, traffic control, barriers, and lights. Table 2.3 is the collection sheet for utilities.

Working Drawing 2.5 shows a storm-drain system, an incoming water line, and a sanitary sewer. The rough grading has been done; the present and finish subgrades are 6 in. below the tops of the manholes and catch basins. The

TABLE 2.3 Collection Sheet for Utilities

Storm sewers

	M.H.		C.B. (+2.5 ft)	6-in. C.P.	10-in. C.P.	12-in. R.C.P.	18-in. R.C.P.
1.	2.85	1.	2.5	44 ft	74 ft	56 ft	80 ft (6.4 ft)
2.	4.00	2.	3.6	20			
3.	5.50	3.	3.0	32			San.
	3)12.35	4.	3.6	52		8-in.	M.H.
	4.12 ft		4)12.7	40		96 ft (6.15 ft)	5.8 ft
			3.2 ft	188 ft			

2 in. water = 99 ft   Present grade 6 in. below finish (top of M.H.)

---

lengths of the pipe runs are shown, as is customary on many engineering site drawings. The site grades are not required, and so have been omitted.

Taking off the excavation of utilities trenches can be a long and tedious process if each run of piping is taken off separately at its exact depth. The storm-drain excavation (W. D. 2.5) will be taken off by two methods, so that we may see how the total quantity obtained by the recommended method compares with the slower and more detailed itemizing of every separate run. The collection sheet is shown above.

The parenthetical notation "(+2.5 ft)" is a reminder that the depths of the catch basins are 2 ft 6 in. greater than the depths listed, which are depth-to-invert figures.

The manholes and catch basins are totaled and averaged. The parenthetical notation (6.4 ft) for the 18-in. reinforced-concrete pipe represents the average depth of the trench; that is, 5.5 ft at manhole No. 3 and 8.3 ft at the street manhole, or an average of 6.9 ft, less the 6 in. that the present subgrade is below the finish grade. The 6.15-ft notation for the 8-in. vitrified clay pipe is the average depth of the trench.

The trench excavation will be taken off by averaging the depths of catch basins and manholes for all the storm drains except the 18-in.:

```
        M.H.      4.12 ft
        C.B.      3.20
                2)7.32
                  3.66
        Less      0.50
                  3.16 ft

              =  3–2 (storm 6-in. to 12-in.)
```

The take-off (W. D. 2.5)

UTILITIES

Trench exc. & backfill

Storm	6-in. to 12-in.	318–0	×	3–3	×	3–2	=	3,273 CF	}	5,839 CF
Storm	18-in.	80–0	×	5–0	×	6–5	=	2,566		
Sanitary		96–0	×	4–6	×	6–2	=	2,664		
Water		99–0	×	3–0	×	3–6	=	1,040		
C.B.	4/	11–0	×	11–0	×	5–9	=	2,783		
M.H.	4/	11–0	×	11–0	×	4–6	=	2,178		
								14,504 CF		
							=	537 CY		

Sheeting util. trenches

$$\left. \begin{array}{c} 2/\quad 80\text{--}0 \times 6\text{--}6 \\ 2/\quad 96\text{--}0 \times 6\text{--}6 \end{array} \right\} \underline{2{,}288 \text{ SF}} \text{ (b.s.m.)}$$

Conc.-pipe storm sewer — 6-in. = <u>188 LF</u>

Conc.-pipe storm sewer — 10-in. = <u>74 LF</u>

Reinf.-conc.-pipe storm sewer — 12-in. = <u>56 LF</u>

Reinf.-conc.-pipe storm sewer — 18-in. = <u>80 LF</u>

8-in. V.-C. pipe sewer = <u>96 LF</u>

Connect 18-in. storm sewer to main = <u>1 ea.</u>

Connect 8-in. san. sewer to main = <u>1 ea.</u>

Street barricades & lights = <u>L. S.</u>

Repair street paving

$$34\text{--}0 \times 9\text{--}0 = \underline{34 \text{ SY}}$$

Permits & fees = <u>L. S.</u>

C. B. 4-0 dia. × *5–9 av. depth (8-in. conc. block)* = <u>4 ea.</u>

M.H. 4-6 dia. × *4–6 av. depth (8-in. com. brick)* = <u>4 ea.</u>

H.-D. C.-I. M.H. cover & fr. 24 in. = <u>4 ea.</u>

H.-D. C.-I. C.B. grated cover & fr. 24 in. = <u>4 ea.</u>

2-in. W.-I. water service = <u>99 LF</u>

Connect water service to city branch main = <u>L. S.</u>

Notes on the take-off (W. D. 2.5)

Compare with the quick, tidy method used above, the take-off of excavation for the storm sewers with each run of sewers taken off at its proper average depth:

6-in.	44–0 × 3–0 × 2.43 ft	=	107 SF	⎫					
	20–0 × 3–0 × 2.20	=	44	⎬ 413 SF × 3–0	=	1,239 CF			
	32–0 × 3–0 × 3.30	=	106						
	52–0 × 3–0 × 3.00	=	156	⎭					
	40–0 × 4–0 × 4.05	=	162			=	648		
10-in.	74–0 × 3–0 × 2.93					=	650		
12-in.	56–0 × 3–6 × 4.25					=	833		
18-in.	80–0 × 5–0 × 6.40					=	2,560		
							5,930 CF		

The difference between the detailed take-off and the "average depth" take-off is only 91 CF, or $3\frac{1}{2}$ CY, which is very small for excavation. That system of averaging the depth must, however, be carefully controlled. Many instances will require a modified method; for example, if there were a considerable number of 4-in. or 6-in. branch lines leading into 8-in. or 12-in. lines, it might be advisable to take off the branch lines as one collected item and the 8-in. and 12-in. lines between manholes as another. The change in depth at drop manholes may require separation of excavation items beyond those points.

The method we used for the trench excavation is quick and simple, and will be accurate in most instances. The 18-in. pipe requires a wider trench than the smaller pipes and so is kept separate. The 12-in. pipe requires a trench of about 3 ft 6 in.; the 10-in. pipe may have a slightly narrower trench; the 6-in. pipe will need a 3-ft trench, which is about as narrow as should be figured for any drainage lines. Thus the average trench is 3 ft 3 in. wide. The first excavation items (combining the 6-in., 10-in., and 12-in. pipe) become 318 ft × 3–3 × 3–2. (Figuring the depth of the trench is explained on p. 41.)

The other trench depths are determined in a similar fashion. The pipe runs are measured between manholes (or between catch basins and manholes); therefore the excavation for manholes and catch basins must be added. That excavation will actually be less than we take off, because in one direction our square hole will overlap 2 or 3 ft of the pipe-trench excavation. The few yards of excavation duplicated is not deducted, however, because it will make up for the extra excavation required if a bank should cave in.

The trenches over 6 ft deep are measured for sheeting. Since one is 6 ft 2 in. deep and the other 6 ft 5 in., 6 ft 6 in. of sheeting is a logical depth to use in both trenches. The height of the walers will probably depend on what material is available at the site. The notation "b.s.m." (both sides measured) will be a consideration when the item is priced.

The pipe items are taken off at net length, in even feet. For pipe sold in 2-ft lengths, this is satisfactory. In a case such as concrete pipe that comes only in 6-ft lengths, the items would be taken off in multiples of 6 ft (for example, 188 LF of 6-in. concrete pipe, would be taken off as 192 LF).

Street connections, barricades, and the like should be checked with the local authorities.

It is not necessary to take off the manholes and catch basins in detail (separating concrete, forms, and brick). Having made two or three estimates using the descriptive method shown in our take-off, an estimator will have worked up a price for a certain depth and a per-lineal-foot price for variations on that basic price, and will thus be able to give a lump-sum price for any manhole of that particular design and diameter.

UTILITIES: SEPTIC TANKS AND DISTRIBUTION BOXES

Working Drawing 2.6 shows a concrete septic tank with cast-iron fittings, brick manholes, and a 5-in. siphon. The piping is shown diagrammatically. The small sketch under W. D. 2.6 shows the natural (and finish) grades in the vicinity of the sanitary system. Working Drawing 2.7 shows the sanitary system extending from the septic tank: the outfall line, distribution box, headers, and field piping.

Starting with the septic tank and the 6-in. length of cast-iron pipe that connects the sewer line to the building (not included and not shown), we shall take off all the items for W. D. 2.6 and 2.7.

The take-off (W. D. 2.6 and 2.7)

UTILITIES — SEWER SYSTEM

Exc. — *septic tank*

$$35\text{--}0 \times 22\text{--}6 \times 12\text{--}7 = 9{,}916 \text{ CF}$$
$$8\text{--}6 \times 22\text{--}6 \times 9\text{--}10 = 1{,}878$$
$$\overline{11{,}794 \text{ CF}}$$
$$= \underline{437 \text{ CY}}$$

Hand exc. — *siphon* = $\underline{1 \text{ CY}}$

Exc. & backfill sewer trenches

$$68\text{--}0 \times 4\text{--}6 \times 6\text{--}6 = 1{,}989 \text{ CF}$$
$$42\text{--}0 \times 4\text{--}6 \times 4\text{--}0 = 756$$
$$+ \text{ Box} \quad 5\text{--}0 \times 4\text{--}0 \times 5\text{--}6 = 110$$
$$\overline{2{,}855 \text{ CF}}$$
$$= \underline{106 \text{ CY}}$$

Excavation and Site Work

Exc. disposal field

$$93\text{-}0 \;\times\; 57\text{-}0 \;\times\; 6\text{-}8 \;=\; 5{,}301 \text{ SF}$$
$$= 35{,}340 \text{ CF}$$
$$= \underline{1{,}310 \text{ CY}}$$

Hand trim for septic tank floor

$$35\text{-}0 \;\times\; 14\text{-}0 \;=\; \underline{490 \text{ SF}}$$

Backfill — septic tank

	Exc.			=	437 CY
Tank 33-8 × 12-6 × 9-1	=	3,824 CF		=	142
					<u>295 CY</u>

Trench sheeting

(part — outfall line)   2/   35-0 × 8-0 = <u>560 SF</u> (b.s.m.)

Pumping = <u>L. S.</u>

Septic tank — conc. (waterproof admix)

                                                      FORMS

Floor
34-8 × 13-6 × 1-0   = 468 CF   110 SF
Walls
81-10 × 1-0 × 7-3   = 594 | 2/   594 SF = 1,188
27-6 × 1-0 × 4-6   = 124 | 2/   124   =   248
Roof
33-8 × 12-6 × 0-10   = 351 | 29-8 × 10-6 =   312
                            1,537 CF | 93-0 × 0-10   =   78
                          = 57 CY | <u>1,936 SF</u>
Add around siphon | 2 × 4 *keyways*
2-0 × 2-0 × 2-9 = 11 CF = $\frac{1}{2}$ | 2 × 110-0   =   <u>220 LF</u>
                         <u>57½ CY</u> | *52-in. dia. hole*
                                       | *through roof* =   <u>2 pcs.</u>

Common brick — M.H.'s over tank

(8 in.)   2/ $3\frac{1}{7}$ × 3-8 × 3-6 = <u>81 SF</u>
                 At $13\frac{1}{2}$ per SF   = 1,094 pcs.
                   + Waste 4%   46
                                    <u>1,140 pcs.</u>

24-in. C.-I. cover & fr. = <u>2 ea.</u>

Galv. step irons (none shown) = <u>20 pcs.</u>

46 Chapter Two

Conc. distr. box

		FORMS

Flr. & roof
$$2/\quad 4\text{-}9 \times 3\text{-}9 \times 0\text{-}6 = 18 \text{ CF}$$
Walls
$$15\text{-}0 \times 0\text{-}6 \times 2\text{-}9 = \underline{22}$$
$$\overline{40 \text{ CF}}$$
$$= \underline{\underline{1\tfrac{1}{2} \text{ CY}}}$$

$$17\text{-}0 \times 3\text{-}9 = 64 \text{ SF}$$
$$\left.\begin{array}{l}13\text{-}0 \times 2\text{-}9 \\ 3\text{-}9 \times 2\text{-}9\end{array}\right\} 46$$
$$\underline{110 \text{ SF}} \left(\begin{array}{c}\text{L.}\\ \$0.90\end{array}\right)$$

6-in. C.-I. sewer pipe = <u>6 LF</u>

5-in. C.-I. sewer pipe
$$6\text{-}0 + 3/\ 2\text{-}0 + 2/\ 2\text{-}0 = \underline{16 \text{ LF}}$$

5-in. C.-I. tees = <u>5 pcs.</u>

6-in. C.-I. tee = <u>1 pc.</u>

6-in. × 4-in. C.-I. tee = <u>1 pc.</u>

Adapter (6-in. C.I. to 6-in. V.C.) = <u>2 pcs.</u>

Adapter (5-in. C.I. to 6-in. V.C.) = <u>1 pc.</u>

5-in. anti-siphon = <u>1 pc.</u>

4-in. C.-I. pipe = <u>6 LF</u>

6-in. V.-C. sewer pipe
$$68\text{-}0 + 54\text{-}0 = \underline{122 \text{ LF}}$$

4-in. V.-C. sewer pipe
$$2\ (16\text{-}0 + 6\text{-}0 + 2\text{-}0 + 2\text{-}0) = 2/\ 26 \text{ ft} = 52 \text{ ft}$$
$$+ \underline{2}$$
$$\underline{54 \text{ LF}}$$

4-in. V.-C. elbows = <u>4 pcs.</u>

6-in. × 4-in. V.-C. tees
$$5 + 8 = \underline{13 \text{ pcs.}}$$

6-in. V.-C. elbows = <u>2 pcs.</u>

4-in. V.-C. pipe disposal field (open joints) = <u>900 LF</u>

Graded gravel bed ($\frac{1}{2}$ in. to $1\frac{1}{2}$ in.)

$$5{,}301 \text{ SF} \times 1\text{--}8 = 8{,}835 \text{ CF} = 327 \text{ CY}$$
$$+ 15\% \quad \underline{49}$$
$$\underline{376 \text{ CY}}$$

Graded gravel bed ($\frac{3}{4}$ in. to $1\frac{1}{2}$ in.)

$$5{,}301 \text{ SF} \times 2\text{--}0 = 10{,}602 \text{ CF} = 393 \text{ CY}$$
$$+ 15\% \quad \underline{59}$$
$$\underline{452 \text{ CY}}$$

9 in. loam (supply and spread)

$$5{,}301 \text{ SF} \times 0\text{--}9 = 3{,}976 \text{ CF} = 148 \text{ CY}$$
$$+ 20\% \quad \underline{30}$$
$$\underline{178 \text{ CY}}$$

Backfill over disposal area (site material)

Exc. = 1,310 CY
Less 327 CY ⎫
      393   ⎬ 868
      148   ⎭
             ─────
             442 CY

Surplus material for removal from site

142 CY
<u>868</u>
1,010
+ Bulkup <u>150</u>
<u>1,160 CY</u>

Clean up & regrade area disturbed = <u>L. S.</u>

Notes on the take-off (W. D. 2.6 and 2.7)

All the excavation items are taken off first, the structures next, then the pipe fittings for the structures, the sewer piping, and finally the disposal field.

The septic tank (Fig. 2.6) is 33–8 × 12–6, the bottom being 12–7 below the present grade. The deep part of the tank is 25–2 long. If we allow about 5–0

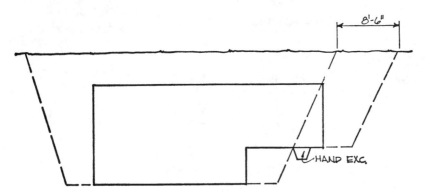

Figure 2.6 Excavation lines for septic tank.

for clearance around the tank, the excavation for the deep section is 35–0 × 22–6. For the siphon area, no working clearance is required in the length, because the 5–0 clearance was taken off in the first item; 1 CY of hand excavation is taken off for around the siphon.

The excavation for the sewer lines includes, as its first item, the outfall line—68–0 from the tank to the distribution box at an average depth of 6–6 (it is 9.35 ft deep at the tank and 3.5 ft at the box). In the second item, 42–0 is the length of the trenching for the lines from the distribution box. The two lines out of the side of the box (one of which is only 6–0 long) both go in one trench; so for excavation there are two lengths at 16–0, plus five 2–0 stubs, or a total of 42–0. Since considerable trenching has been taken off all around the box, the extra excavation for the box is taken off at about its net size. The box is 18 in. deeper than the pipe invert and consequently its excavation is 18 in. deeper (that is, 5 ft 6 in. instead of 4 ft). Notice that the header line in the disposal field is not measured for trench excavation. The disposal-field excavation will be 1 ft 6 in. outside the pipe lines to provide sufficient room for the header line, which is part of the field piping.

For a concrete slab such as the septic-tank floor, some hand labor is always needed for clearing out loose dirt, leveling, and the like; therefore a handtrim item is taken off.

The backfill at the septic tank is taken off by simply deducting the volume of the structure from the total excavation item. If we consider the structure as a simple box 33–8 × 12–6 × 9–1, the small excess volume taken where the siphon chamber is 6–5 deep will offset the minor items not deducted (the two manholes, the concrete around the siphon, and the 6-in. projection of the tank bottom slab).

The trench sheeting item includes about half of the 68-ft line. That line is 9.35 ft deep at one end and 3.5 ft deep at the other, so that at about 35 ft from the septic tank the excavation will become less than 6 ft deep; sheeting is not taken off beyond that point. The 8-ft depth of the sheeting is an average for the walers, which will vary from 10 to 6 ft in height. Notice the "b.s.m." notation, indicating "both sides measured."

Pumping is taken off, because although the water level is below the deepest excavation grade, surface water can always be a problem in bad weather.

The concrete and formwork for the septic tank are taken off as single items. The two items are sufficient for such a small structure, because overlapping job costs make it difficult to separate or price the items in greater detail. The openings through the roof are not deducted. The item for roof forms consists of the inside surface area (the width—10 ft 6 in.—times the sum of the lengths of the 3 chambers) plus the area of the 10-in. edge forms (the outside perimeter times 10 in.).

The concrete around the siphon was missed; picked up later, it is shown as an "add" to the total for the septic tank.

The manhole is taken off—at its mean diameter of 3 ft 8 in.—"twice times," for two manholes. The 8-in. brick is taken off at $13\frac{1}{2}$ bricks per SF—a factor that is about right for common-brick work in minor items. (The entire question of brick quantities is discussed in Chap. 6.)

Step irons have been taken off, although they are not shown. Reference to the specifications may clarify an item such as this.

The distribution box is a straightforward take-off, except that although the top slab is not shown as such, it would have to be precast. With this in mind, a note is made $"\begin{pmatrix} L \\ \$0.90 \end{pmatrix}"$ as a reminder that this small item is worth $0.90 per SF for formwork labor, and that $0.90 will be entered in the labor unit-price column of the estimate when the take-off item is entered.

The cast-iron pipe is measured in 2-ft multiples for the short pieces that are connected to the tees inside the septic tank. Lengths of cast-iron pipe that have a collar are measured in 6-ft multiples. The 6-in. cast-iron tee is for inside the distribution box; the adapters are to join the cast-iron pipe to the vitrified clay pipe.

The 4-in. vitrified clay pipe specials are for the bends in the four lines from the distribution box. The 6-in. $\times$ 4-in. tees connect the distribution lines to the header line (5 tees), and the header line to the field lines (8 tees). A 6-in. elbow is needed for each end of the header line.

The gravel for the disposal field has 15 per cent added for compaction and waste; the topsoil has a 20 per cent allowance. The surplus material for the disposal item is the sum of the septic tank displacement and the disposal field displacement.

MISCELLANEOUS UTILITIES ITEMS

Precast concrete septic tanks are taken off as a lump sum in which case a unit price should be obtained from suppliers. The excavation, setting, and piping must be taken off separately.

Cesspools may be described fully for lump-sum pricing of labor and material, or they may be taken off item by item.

Drain inlets are described and the average depth stated for lump-sum pricing. This method also applies to manholes and catch basins.

Headwalls for storm-drain outfalls are taken off in detail to include excavation, concrete, forms, and stone-bed.

Dry wells are taken off in the same way as cesspools.

When pricing sewer or drain piping, check on the material specified. Some corrugated culvert pipe comes only in long lengths. Some large-diameter sewer pipe is very heavy and must be unloaded and placed by a crane, or even by two cranes. Heavy sewer piping (24 in. and larger) should be considered as material requiring special handling. The take-off must provide for such handling. Also consider the costs of engineering, excavation, unloading and distributing the pipe, method of laying, shoring the trench, and the cost of pipe cut to length.

Footing drains are taken off in the same way as other drainage items, except that if the footing drains are around the perimeter of a building, the pipe usually cannot simply be dropped into the working area excavated for the foundation walls. Since the grade of the pipe will vary, some hand excavation will be required, and must be included in the take-off. If the footing drain is to be surrounded by porous gravel, forms may have to be taken off for the outside face of the surround. This is a small but expensive operation, involving the gradual withdrawal of the form as the outside backfill is placed in order not to disturb the gravel around the drain.

Dewatering. Water can become a major problem and may have to be handled as a special item. If there is a definite water condition that will affect the work, all aspects of the condition must be carefully examined. A well-point company may have to be consulted. Generally, it is not the policy of well-point companies to give firm prices for dewatering. They will advise on the amount of equipment required and give rental rates for the various pieces of equipment, but usually the general contractor will be responsible for finally evaluating and pricing the item.

Every job of this sort must be considered as a special case presenting its own problems; there is no common approach. But one important question should always be asked: when were the borings taken and what has the weather been like since then? A long dry spell could have lowered the water level considerably. If it will be necessary to divert water, temporary drainage ditches should be measured and taken off as "Exc. temp. drainage ditch—...CY"; if the diversion is to be effected by damming the ditch and pumping, then items should be taken off for that work and properly described.

Sundry site work items

Fieldstone paving is measured in square feet and fully described. The sand or concrete bed for paving is taken off separately from the stone paving. Edge forms for pavings are taken off in lineal feet, and the size described.

Steel edging for lawns and the like is taken off in lineal feet and described. The price obtained from the manufacturer will usually include the stakes.

Concrete curbing if precast may be a standard manufactured item, for which a price can be obtained. If that is so, take it off per lineal foot with separate items for excavation and the like. Cast-in-place concrete curbing must be taken off in several separate items, such as excavation, concrete, forms, rubbing, and backfilling.

Fence postholes and concrete for setting the posts are taken off as one item: "Concrete surround for fence posts, incl. exc.—29 ea." Ordinary fence postholes without concrete are not taken off—both the hole and the setting are included in the post item.

Wood fences are taken off in several items, as for example: "3-in. × 4-in. creosoted fence posts 8 ft long—...pcs.; 2 × 4 creosoted fence rails—...BF; and 1-in. × 3-in. cedar fencing 5 ft long, pointed—...pcs." Boarding for close-boarded fencing is taken off in board feet; chestnut fencing in lineal feet of stated height.

Chain-link fencing is measured in square feet or lineal feet of the specified type, gauge, and height, with the extra items, such as posts, stays, and gates, taken off separately.

Retaining walls of concrete are included in Chap. 3. Stone retaining walls are taken off in cubic feet and the type of stone described.

Seats are usually treated as a separate item, fully described for a lump-sum unit price, if they are a standard manufactured item. Specially designed seats are taken off item by item.

Flagpole bases may be taken off in several separately priced items, such as excavation, concrete, forms, and setting of the sleeve. Having priced a few jobs including flagpole bases, the estimator may reduce this item to a single unit price; for example, "3-ft dia. × 4 ft deep flagpole base—L. S."

Baseball fields require, in addition to the grading and seeding (which are taken off in the usual manner), several special items. Items such as excavate, base course (gravel or as specified), and clay and sand (1 to 1) top course (all for skinned area) are taken off in cubic yards; fine grading (skinned area) is taken off in square feet; form pitcher's mound is a lump-sum item. Exact details would be as specified.

Cinder running tracks require the following items, taken off in cubic yards: excavation (or fill), stone base course (as specified), cinder bed, and surface course (usually brown clay and cinders mixed 1 to 2). In addition, an item for roll and shape subbase and another for fine grade and roll are taken off in square feet. The entire take-off for athletic fields will follow the requirements of the specifications. Some running tracks require special latex or rubber-based asphalt paving. Prices for this work should be obtained from subcontractors.

SITE WORK: DEEP PIT IN WATER

Working Drawing 2.8 shows a concrete pit 16 ft deep inside, with the bottom of the pit about 8 ft below the level of the groundwater. The drawings may not show the water level. The estimator should check the borings to determine the water level. The decision to use temporary sheet piling to support the excavation and help control the water would also be made by the estimator. Note that the references to water level and sheet piling on drawing 2.8 were made by the estimator.

Chapter Two

The take-off (W. D. 2.8)

Exc. machine, Stage 1

$$24\text{-}0 \times 24\text{-}0 \times 6\text{-}6 = 3{,}744 \text{ CF}$$
$$= \underline{139 \text{ CY}}$$

Steel sheet piling 27#/SF (drive & pull)

$$4 \times 16\text{-}0 \times 18\text{-}0 = \underline{1{,}152 \text{ SF}}$$

Temp. bracing 16 ft × 16 ft × 12 ft deep = $\underline{\text{L. S.}}$

Exc. inside sheet piling

$$16\text{-}0 \times 16\text{-}0 \times 10\text{-}6 = 2{,}688 \text{ CF}$$
$$= \underline{100 \text{ CY}}$$

Exc., hand, bottom of floor

$$45\text{-}0 \times 3\text{-}3 \times -6 = 74 \text{ CF}$$
$$= \underline{3 \text{ CY}}$$

Hand-trim bottom

$$15\text{-}0 \times 15\text{-}0 = \underline{225 \text{ SF}}$$

Pumping pit (4 weeks) $\underline{\text{L. S.}}$

Concrete 3,000-lb bottom slab

				Formwork		
45-0 × 3-3 ×	-6 =	74 CF	4 × 8-0 ×	-6 =	16 SF	
14-6 × 14-6 ×	1-0	351	4 × 14-6 ×	1-6	87	
2-6 × 2-0 ×	1-0	5	9-0 ×	1-6	14	
+ 1-6 × 2-0 ×	-6	2			$\underline{118 \text{ SF}}$	
		432				
	=	$\underline{16 \text{ CY}}$				

Concrete, walls

$$45\text{-}0 \times 1\text{-}3 \times 16\text{-}0 = 900 \text{ CF} \qquad 2 \times 720 \text{ SF} = \underline{1{,}440 \text{ SF}}$$
$$= \underline{33.5 \text{ CY}}$$

Concrete roof

$$12\text{-}6 \times 12\text{-}6 \times 1\text{-}0 = 157 \text{ CF} \qquad 10\text{-}0 \times 10\text{-}0 = 100 \text{ SF}$$
$$\qquad\qquad\qquad\qquad\qquad\qquad\qquad 45\text{-}0 \times 1\text{-}0 = 45$$
$$\qquad\qquad\qquad\qquad\qquad\qquad\qquad\qquad\qquad\qquad \underline{145 \text{ SF}}$$
$$= \underline{6 \text{ CY}} \qquad 2 \times 4 \text{ Keyway} = \underline{90 \text{ LF}}$$

Float-finish slabs

$$\begin{array}{rcl} 10\text{-}0 \times 10\text{-}0 & = & 100 \\ 12\text{-}6 \times 12\text{-}6 & & \underline{156} \\ & & \underline{256 \text{ SF}} \end{array}$$

C.I. cover & frame 24 × 24 × 87 lb. = <u>1 Set</u>

Steel Ladder $\frac{1}{4}$ in. × $1\frac{1}{4}$ in. with $\frac{3}{4}$-in. dia. rungs

@ 12 in. c-c, 14 in. wide = <u>15 LF</u>

$\frac{3}{4}$ in. × 18 in. Bolt + nut & washer = <u>4 ea.</u>

Reinforcing steel

								No. 4	No. 5
Bot. slab	No. 4	2	×	8	×	14 ft	=	224	
	No. 5	2	×	4	×	14 ft			112 ft
	No. 5	2	×	4	×	8 ft			64
Bot. hooked	No. 4	4	×	12	×	5 ft		240	
Walls	No. 4	4	×	16	×	12 ft		768	
	No. 4	4	×	12	×	16 ft		768	
Top hooked	No. 5	4	×	12	×	5 ft			240
Roof	No. 5	2	×	13	×	12 ft			312
	No. 5	2	×	3	×	10 ft			60
								2,000 ft	788 ft
	No. 4	2,000	×	0.668	=			1,336 lb	
	No. 5	788	×	1.043				823	
								2,159	1.1 ton

Backfill, compacted

$$\begin{array}{lrcl} \text{Total excavated} & & = & 239 \text{ CY} \\ \text{Less:}\ 10\text{-}0 \times 10\text{-}0 \times 16\text{-}0 = 1{,}600 \text{ CF} & = & & 59 \text{ CY} \\ \text{Total concrete} & & & \underline{55} \\ & & & 114 \\ & & & 114 \\ & & = & \underline{125 \text{ CY}} \end{array}$$

PVC Waterstop $\frac{3}{16}$ × 6 dumbbell = <u>45 LF</u>

Notes on the take-off (W. D. 2.8)

Open excavation is taken for the first 6 ft 6 in. of depth, which is to within 2 ft of the waterline. The average width of this square hole is 24 ft. This is cheaper than driving the sheet piling from the ground level. Next, the interlocking steel sheet piling would be driven. This is not precision work; therefore, 18-ft lengths of piling are figured compared to the theoretical 16-ft lengths. Although the piling has been measured, a subbid would probably be solicited for the item for comparison.

Note that the take-off does not attempt to detail the bracing or the dewatering. The methods of performing these items need thought and will be figured out when the estimate is priced.

The more costly excavation inside the sheet piling is kept separate from the open excavation. Items for handwork in the bottom of the hole are also taken off.

For clarity, the concrete is shown in three separate parts, floor, walls, and roof. An experienced estimator may elect to combine these into a single item and price it accordingly. In the bottom slab, the thickest section is taken first. The computation of the length could be done either by adding the two outside lengths to the two inside lengths or by using the mean dimension of the pit at the center of the walls:

Out to out $2(1\text{--}0 + 1\text{--}3 + 10\text{--}0 + 1\text{--}3 + 1\text{--}0) = 2 \times 14\text{--}6 = 29\text{--}0$
$\qquad\qquad 2[14\text{--}6 - (2 \times 3\text{--}3)] \qquad\qquad\quad = 2 \times 8\text{--}0 = 16\text{--}0$
$\qquad\qquad\qquad\qquad\qquad\qquad\qquad\qquad\qquad\qquad\qquad\qquad\ \ 45\text{--}0$

Mean $\quad 4(10\text{--}0 + 1\text{--}3) \qquad\qquad\qquad\qquad = 4 \times 11\text{--}3 = 45\text{--}0$

Everything being symmetrical, the wall perimeter is the same as the footing perimeter.

The reinforcing steel is measured in even foot lengths and reduced to tonnage. The reinforcing may be bid by a sub, as may be the piling and the miscellaneous iron. All items for the pit have been taken off in this example so as to show how the various trades are handled. It is site-work items such as these that subs may miss or exclude. It is easier to adjust the subbid on bid day than it is to start taking off the items not covered.

The waterstop was missed when it should have been taken off (after the concrete bottom slab), so it was added at the end. It is not important to do so, but in writing up the estimate, this item could be inserted in a more logical place.

Chapter 3

Concrete

In taking off quantities for a building, it is customary to start with the foundation concrete; therefore, the exterior-wall perimeter should be one of the first computations made.

WALL PERIMETER

The rules for computing building exterior-wall perimeters are very simple. In any rectangular building (either a true rectangle or a rectangular design with corners that step in and out) the wall perimeter is twice the sum of the length and the width, expressed as

$$2(L + W)$$

In making this computation, the four corners must be deducted to allow for the overlap; therefore one outside and one inside dimension should be used in the computation. For Building A in Fig. 3.1, for example:

$$\text{Wall perimeter} = 2(23\text{-}0 + 19\text{-}0) = 84 \text{ LF}$$

It should be noted that 19–0 is 21–0 less 2 times 12 in., the wall thickness.

In the drawing of Building B, it is obvious that the wall lines of the two indented steps are the same length as the dotted lines that superimpose the outline of Building A on it. Thus Building B, having outside dimensions of 23–0 × 21–0, must have a wall perimeter equal to that of Building A (84 LF). If walls should step in as a recess in the middle of one side of a building, however, twice the depth of the recess must be added to the total of 2 times the sum of L and W.

In the working example (W. D. 3.1), the wall perimeter is computed as follows:

		79– 5
		11– 2
Outside	=	90– 7
Inside	=	58– 4
		148–11
	+	148–11
+ Recess	=	8– 6
+ Recess	=	8– 6
Wall perimeter		314–10

No matter how often a wall perimeter line steps in or out, there are only four corners to deduct in computing it. This rule may be expressed as: The wall perimeter is the sum of the outside dimensions less four times the wall thickness.

After the wall perimeter is obtained, that figure can be adjusted to obtain other items. For W. D. 3.1, the outside-wall-surface perimeter will be 314–10 plus 4 times 1–0, or 318–10. The brick-shelf perimeter will differ from the wall perimeter, because the 5-in. shelf is a deduction from the outer face of the 12-in. wall; therefore the brick shelf is 7 in. longer than the wall at each of the four outside corners. Thus for W. D. 3.1, the total brick shelf is 314-10 plus 4 times 0–7, or 317–2.

These perimeter rules can be applied to any building; for example, Fig. 3.2. (The dimensions not shown have been omitted because they are not required for the perimeter computation.)

The wall perimeter for a circular building would be computed using the diameter to the center of the exterior wall.

It should only be necessary to compute the perimeter of a building once—you should then make that figure work for you.

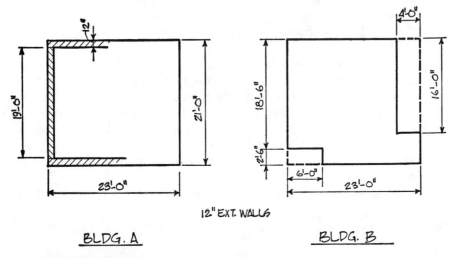

Figure 3.1 Building perimeter.

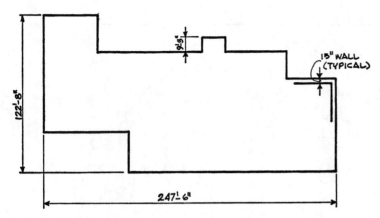

Figure 3.2 Building perimeter.

```
Length (outside)    =   247- 6
Width  (inside)     =   120- 2
                        ------
                        367- 8
                        367- 8
+  2  ×  9-3        =    18- 6
                        ------
Wall perimeter          753-10
```

CONCRETE FOUNDATIONS

Working Drawing 3.1 shows a sketch of the concrete foundations for a building with 12-in. foundation walls, a brick shelf, a section of 16-in. wall, and wall and interior footings. Each interior footing has a 12-in. × 12-in. pier. Proceeding in a predetermined, orderly manner, the foundation concrete is taken off in the following order: pier footings; piers; wall footings; walls; ground slabs (except that the walls are used to obtain the wall footings).

Concrete (3,000 psi unless noted)

Conc. pier ftgs.

									FORMS
A	7/	3-0	×	3-0	×	1-2	=	74 CF	98 SF
B	3/	4-6	×	3-0	×	1-4	=	54	60
C	2/	5-0	×	4-0	×	2-0	=	80	72
								208 CF	230 SF
							=	8 CY	

Having completed this first item, which was simple enough to be entered directly on the take-off sheet, it is time to use a collection sheet, which is shown in Table 3.1.

The collection sheet may be followed through item by item (taking off the

TABLE 3.1 Collection Sheet for Foundation Walls

Foundation walls

				5-in.			
	12-in.		16-in.	brick shelf	Piers		
6–3	7–3	8–3	7–3	1–4½	12 in. × 12 in.		
49–11	29– 6	25–0	19–6	90–7	(A) 2/ 6– 9	=	13– 6
28– 4	33– 4	40–5		59–6	(A) 1/ 7– 9		7– 9
11– 2	13– 8	65–5		150–1	(A) 4/ 5– 9		23– 0
17– 8	8– 6			8–6	(B) 1/ 7– 7		7– 7
107– 1	8– 6			8–6	(B) 2/ 5– 7		11– 2
	17– 0			317–2	(C) 1/ 5–11		5–11
	12– 4				(C) 1/ 4–11		4–11
	122–10				(12)		73–10

Wall ftgs.

24 × 12   28 × 12
295–4   19–6

walls first, the items started at the lower left-hand corner of the figure and proceeded clockwise).

The top of the wall is at a grade of 101–3 and the bottom of the 12-in. footing at 94–0; therefore the wall height is 6–3 for a distance of 49–11, until it steps down to elevation 93–0 for a 7–3 wall 29–6 long. The 79–5 side gives both these first two items. The next wall must be taken off at its inside dimension—58–4—which, less 25–0 at a height of 8–3 (grade 92–0), leaves 33–4 at 7–3. This method is further applied in taking off the remaining portions of the perimeter.

The 5-in. brick shelf is a straightforward perimeter addition: the outside dimension one way (90–7) and inside dimension the other way (60–4 less 0–10), plus the 8–6 side of the recess, all doubled.

The piers can be followed using top and bottom grades, starting with the A piers. For simplicity these are not listed clockwise; they start with the top line (A footing—pier bottom at grade of 94–2, top of pier at 100–11, a height of 6–9, quantity of 2).

The footings follow the walls: the 16-in. wall has a 28-in. × 12-in. footing, 19–6 long; the length of the 24-in. × 12-in. footing is the sum of all the 12-in. walls (107–1 plus 122–10 plus 65–5).

The total length of the exterior walls (which is also the footing length) is 314–10, which checks with our previous calculation of the perimeter.

After these items are transferred from the collection sheet, the take-off sheet will read as follows:

The take-off (W. D. 3.1)

Wall ftgs.

$$
\begin{array}{r}
295{-}\ 4\ \times\ 2{-}0\ \times\ 1{-}0\ =\ 591\ \ \text{CF} \\
\underline{19{-}\ 6}\ \times\ 2{-}4\ \times\ 1{-}0\ =\ \underline{\ \ 46\ \ } \\
314{-}10\hspace{3.2cm}637\ \ \text{CF} \\
=\ \underline{23\tfrac{1}{2}\ \text{CY}}
\end{array}
$$

$\Big\}$ $\underline{630\ \text{SF}}$

$2\ \times\ 4$ *footing*
     keyway $=\ \underline{315\ \text{LF}}$

Foundation piers

$$(12)\quad 1{-}0\ \times\ 1{-}0\ \times\ 73{-}10\ =\ \underline{3\ \text{CY}}$$

$\underline{296\ \ \text{SF}}$
(12 piers,
12 in. × 12 in.)

Conc. ext. fndtn. walls

FORMS

12-in.
$\begin{array}{l}
107{-}\ 1\ \times\ 6{-}3\ =\ \ \ 669\ \text{SF} \\
122{-}10\ \times\ 7{-}3\ =\ \ \ 892 \\
\ 65{-}\ 5\ \times\ 8{-}3\ =\ \ \ 530
\end{array}\Big\}\ 2{,}091\ \text{SF}\ =\ 2{,}091\ \text{CF}$

16-in.
$\begin{array}{l}
\underline{\ 19{-}\ 6}\ \times\ 7{-}3\ =\ \underline{\ \ 141} \hspace{2cm} =\ \ \underline{\ \ 188} \\
314{-}10 \hspace{2.2cm} 2{,}232\ \text{SF} \hspace{1cm} 2{,}279
\end{array}$

Less brick shelf
$317{-}2\ \times\ 0{-}5\ \times\ 1{-}4\tfrac{1}{2} \hspace{2cm} =\ \underline{\ \ 182} $
$\hspace{7cm} =\ \underline{2{,}097}\ \text{CF}$
$\hspace{7cm} =\ \underline{\ \ 78\ \text{CY}}$

$2\ \times\ 2{,}232\ \text{SF}\ =$
$\hspace{2cm}\underline{4{,}464\ \text{SF}}$

5-in. *brick shelf* $=\ \underline{436\ \text{SF}}$

1-in. *chamfer strip to walls*
$\hspace{3cm}=\ \underline{320\ \text{LF}}$

12-in. gravel for ground slab

$$
\begin{array}{r}
88{-}7\ \times\ 58{-}4 \hspace{2cm} =\ 5{,}167\ \text{SF} \\
\text{Less}\quad 18{-}8\ \times\ 11{-}2\ =\ 208 \\
21{-}6\ \times\ \ 8{-}6\ =\ 182
\end{array}\Big\}\ \ \ \ \underline{\ \ 390}
$$

$\hspace{6cm} 4{,}777\ \text{SF}$
$\hspace{4cm} \times\ 12\ \text{in.}\ =\ \underline{4{,}777}\ \text{CF}$

$\hspace{7cm} =\ \ 177\ \text{CY}$
$+\ \text{Waste \& compaction}\ 15\%\ \ =\ \ \ \ 26$
$\hspace{7cm}\underline{203\ \text{CY}}$

6-in. × 6-in. × 10/10 wire mesh

$$
\begin{array}{r}
4{,}777 \text{ SF} \\
+ \text{ Waste } 10\% \quad \underline{473} \\
5{,}250 \text{ SF}
\end{array}
$$

Concrete (2,500 psi) ground slab

4 in. × 4,777 SF = 1,592 CF
= $\underline{59 \text{ CY}}$

Screeds for 4-in. slab = $\underline{480 \text{ LF}}$

Cure slab = $\underline{4{,}777 \text{ SF}}$

Trowel finish floor = $\underline{4{,}777 \text{ SF}}$

½-in. × 4-in. premolded expansion jointing = $\underline{320 \text{ LF}}$

Carbo. rub exterior walls

319–0 × 1–3 = $\underline{400 \text{ SF}}$

Notes on the take-off (W. D. 3.1)

Being all 12 × 12, the foundation piers were added to make one item, but the number of piers (12) is noted; on the estimate the item will read "Concrete foundation piers (12)—3 CY." Note that the final figure was rounded out. Concrete items should be rounded off to the nearest cubic yard if the item is over 50 CY. On the estimate sheet, the formwork for this item is also noted with the number of piers. It is unnecessary to waste time writing the details for the formwork computations, as the concrete items show all that information.

The concrete side of the wall footings item is self-explanatory. As there is only one depth of footing, the formwork is simply 2 times 315 SF, or 630 SF. This item is an "automatic" mental calculation, so the total is inserted without unnecessary detail. The keyway is shown on the drawing and is the same length as the wall footings.

Foundation walls of uniform thickness are kept together and the surface areas extended and totaled, so that the total volume can then be obtained by a single operation. Since these walls are 1 ft thick, no multiplication is necessary to convert square feet to cubic feet; with any other wall thickness, however, it is advantageous to total the areas before multiplying by the thickness. The brick shelf is deducted from the concrete total, but added (as an extra-cost item) to the formwork side; this takes care of the double forming required at the brick shelf. Various extra-cost items for walls (such as pilasters, slab keys, and the like) are handled later (pp. 77 and 78).

The gravel for the slab item is the ground-slab area; note that the isolated piers or pier footings are not deducted. The required quantity of gravel cannot be computed exactly; truck measure is not itself an exact quantity, and the compaction varies according to the grading of material, its moisture content, and the percentage of compaction required. All factors considered, however, 15 percent is a reasonable allowance for spillage, waste, and compaction, although sometimes it might be necessary to allow 20 percent.

The wire mesh item is also the ground-slab area, with an additional 10 percent for laps and waste. For large areas, the waste and laps may be reduced to 8 percent; for small, cut-up areas, 12 percent may be needed. Note that in adding the waste, the figure used is adjusted slightly so as to round off the total quantity (10 percent of 4,777 SF is 478 SF, which is decreased by 5 SF to 473 SF, to round off the total to 5,250 SF).

The floor-slab item is straightforward. The formwork for screeds is one of those items that is needed but does not show on the drawings. Screeds should be figured at about 1 LF of screed to every 10 SF of floor, which assumes the use of a 12-ft to 14-ft straight edge, with the screeds set on between 10-ft and 12-ft centers.

If specified, the type of curing should be described. If burlap covering is called for, take it off and add 10 percent waste: "Burlap for curing slab, sealed joints, 3 days wetted—5,250 SF."

The trowel finish item is straightforward. Normally the concrete finishing items are not taken off until all the structural items have been taken off. The finish items are then taken off from the floor finish schedule and added to the collection sheet. For floors, measure the areas that differ from the predominant finish and deduct their total area from the gross floor area to obtain the area of the predominant finish.

The expansion joints were not mentioned on the plans, but on the section there is a dark line between the floor and the wall; you would go to your specifications takeoff for a description of the item. The quantity is the building perimeter with a 10 percent waste allowance added.

Although only 12 in. of wall is to be exposed, the finish grade could hardly be expected to be on a perfect line; thus, the rubbing item should extend a little lower than the required line. At least a 4-in. extra band should always be allowed.

This completes the take-off of items that are either shown in W. D. 3.1 or can reasonably be assumed to be included in that drawing. The specifications take-off should determine items such as expansion jointing, rubbing, curing, and trowel finish.

AREAWAY AND STAIRS

Figure 3.3 shows a basement entrance. The main wall of the building is not to be included in this example.

62 Chapter Three

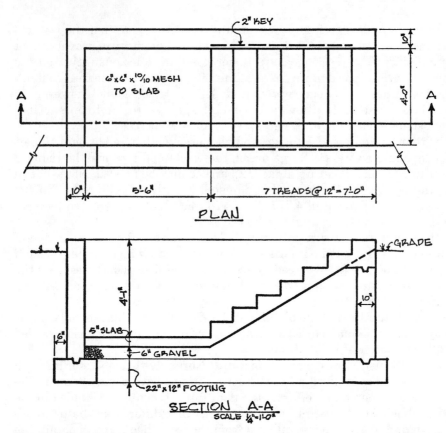

Figure 3.3 Basement stairs.

The take-off (Fig. 3.3)

Conc. wall footings.

$$21\text{-}4 \times 1\text{-}10 \times 1\text{-}0 = \underline{\underline{39}} \text{ CF}$$
$$= \underline{\underline{1\tfrac{1}{2}}} \text{ CY}$$

FORMS
43 SF
2×4 *ftg. key* = 22 LF

Conc. foundation walls

$$17\text{-}4 \times 0\text{-}10 \times 5\text{-} 0 = 87 \text{ SF}$$
$$4\text{-}2 \times 0\text{-}10 \times 3\text{-}10 = \underline{16}$$
$$\underline{\underline{103}} \text{ SF} = 86 \text{ CF}$$
$$= \underline{\underline{3\tfrac{1}{4}}} \text{ CY}$$

206 SF
Wall key = 4 LF

Conc. — bast. stair and slab

Slab
$$5\text{-}6 \times 4\text{-}0 = 22 \text{ SF} \times 5 \text{ in.} = 9 \text{ CF}$$

Stair
$$8\text{-}6 \times 4\text{-}0 \times 1\text{-}0 = \underline{34}$$
$$\underline{43} \text{ CF}$$
$$= \underline{\underline{1\tfrac{1}{2}}} \text{ CY}$$

$8\text{-}6 \times 4\text{-}0$
$4\text{-}1 \times 4\text{-}0$ } 51 SF

6-in. × 6-in. × 10/10 mesh   =   25 SF

Float or rub, steps and slab

$$\left.\begin{array}{l}12\text{-}6 \times 4\text{-}0 \\ 4\text{-}1 \times 4\text{-}0\end{array}\right\}\ 67\ \text{SF}$$

Rub walls

$$\left.\begin{array}{ll}\text{Int.} & 22\text{-}0 \times 4\text{-}1 \\ \text{Ext.} & 23\text{-}0 \times 1\text{-}0\end{array}\right\}\ 113\ \text{SF}$$

Gravel bed for areaway   =   ½ CY

Notes on the take-off (Fig. 3.3)

For an actual building, all the areaways would be taken off together, so the individual items would show larger quantities than the above example; also the figures would probably be rounded off to the nearest cubic yard. For a building that had only one areaway (as the above example), all the concrete could be taken off in one item, as follows:

Conc.—areaways, stairs

		FORMS
Ftg. 21-4 × 1-10 × 1-0 = 39 CF		43 SF
Walls		
$\left.\begin{array}{l}17\text{-}4 \times 0\text{-}10 \times 5\text{-}0 = 87 \\ 4\text{-}2 \times 0\text{-}10 \times 3\text{-}10 = 16\end{array}\right\}\ 103$ = 86		206
Slab 22 SF × 0-5 = 9		
Stair 8-6 × 4-0 × 1-0 = 34		51
168 CF		300 SF
6 CY		2 × 4 keyways = 26 LF

The stair is taken off by measuring the soffit (8 ft 6 in.) and multiplying it by the width (4 ft) and the thickness (1 ft). Although the actual thickness of the stairway may be 9 or 10 in., it is advisable to use 12 in. for all ordinary steps, to allow for the high percentage of waste in pouring. The stair form-work item is the "contact area": the area of the soffit plus that of the risers. For an open stair, the area of the sides would be added in the formwork. If the walls are to be poured and stripped, and the area behind the stairs back-filled and graded off to the stair slope without using forms, the cost will be about the same as would be carried for the forms; therefore one may simply carry soffit forms rather than backfill and "Hand trim of slope."

The rubbing item includes the three sides of the entrance slab section (5–6 plus 4–0 plus 5–6) plus half the wall at the stairs (2 sides at 3–6 each) for a total of 22–0 in length—4–1 high. Note that the rubbing of the main building

64   Chapter Three

wall is included in this item. The finishing of the slab, risers, and treads is carried as a single total. It would be impossible to separate the cement finisher's time into three individual items of cost.

EQUIPMENT PADS

Take-off of equipment pads in Fig 3.4

Conc. eqpt. pads

8/ 8–8 × 3–0 × 1–0	=	208 CF
8/ 4–2 × 3–3 × 1–0	=	108
		316 CF
	=	12 CY

Trowel finish eqpt. pads = 316 SF

Rub sides eqpt. pads = 120 SF

FORMS

8 × 30 ft = 240 LF
(12-in. edge form)

3-in. chamfer strips = 240 LF

$\frac{1}{2}$-in. × 6-in. exp. jointing = 240 LF

Notes on the take-off (Fig. 3.4)

The formwork is taken off in lineal feet rather than square feet; the estimate sheet will read: "12-in. edge forms for equipment pads—240 LF." This method is used for items involving edge forms up to 12 in. high; pads, curbs, and bases over 12 in. high should be taken off in square feet. There is no concrete deduction for the chamfer strip.

The rubbing and the trowel finish could be lumped together, providing a suitable unit price is used. If rubbing is worth $0.17 per SF (for 120 SF) and trowel finish $0.12 per SF (for 316 SF), then the combined item ("Finishing concrete pads") would be priced at $0.14 per SF—the average price of the two items after rounding off.

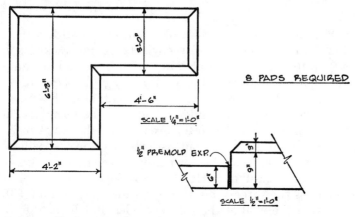

Figure 3.4 Equipment pads.

Take-off of concrete floor pipe trenches (W. D. 3.2*)

Conc. ground slab (5 in.)

54–6 × 48–8 =	2,652 SF		*FORMS*
23–0 × 12–2 =	280		5-in. edge form = 98 LF
	2,932 SF	2,932 SF	

Less trench C-C   29–6 × 4–4 = 128
Less trench C-C   39–6 × 4–2 = 165 } 320   *Trench B-B*
Less trench B-B   17–0 × 1–0 = 27
                          2,612 SF   59–0 × 1–1 = 64 SF
                                       57–0 × 1–0 = 57
5-in. slabs   = 1,088 CF   121 SF
+ B-B   28–3 × 2–0 × 0–6 =   28   *5-in. slab screeds* = 300 LF
         58–6 × 0–6 × 0–7 =   17
                             1,133 CF
                          =   42 CY   *Set angle-iron frame* = 58 LF

Pipe trench C-C

  Conc. wall ftg.

       98–2 × 1–4 × 0–8 = 88 CF   132 SF
                        = 3½ CY   *Ftg. key* = 99 LF

  Conc. pipe trench walls

  8 in. 54– 8 × 2–7 = 141 SF
         38–10 × 3–7 = 139   604 SF
          4– 8 × 4–8 = 22
                    302 SF × 8 in. = 202 CF
                                 = 7½ CY

  Conc. — pipe trench & pit floor

    39–10 × 3–8 = 146 SF
    29– 0 × 3–4 = 97
                 243 SF × 3 in. = 61 CF
   + Drop 3–4 × 0–6 × 1–0   }   4   *12-in. edge forms* = 8 LF
           3–8 × 0–6 × 1–1
                                65 CF
                          = 2½ CY

* The small trench (Sec. B-B on W. D. 3.2) is taken off with the floor slab.

Conc. slab over pipe trench

65–0 × 4–4 = 282 SF × 5 in. = 118 CF | 260 SF (note stripping)
= 4½ CY

6-in. × 6-in. × 10/10 *mesh for slab* = 270 SF

Notes on the take-off (W. D. 3.2)

All parts poured with the slab are taken off with it as part of the slab concrete item. On the formwork side, however, the various items are taken off separately. The top slab of a pipe trench such as Sec. C-C is often a continuous part of the floor; in that case, the entire floor slab would be taken off as one item with the formwork for the slab over the trench described and measured.

In the 4-ft-wide pipe trench, the wall of the pit has been taken off at 4 ft 8 in. deep from the underside of the floor, although in the drawing the pit could be 4 ft 8 in. from the top of the slab. The difference would be very slight.

The "drop" items added to the Sec. C-C type pit floor are for the changes in floor grade (Fig. 3.5), for which the 12-in. edge forms are also taken off.

The notation "(note stripping)" against the pipe-trench top slab item is a reminder to the estimator that stripping forms from this low trench will be expensive.

Precast concrete trench covers are taken off by pieces and described. Consider Sec. C-C (W. D. 3.2), with precast covers in 4-ft lengths, 3 in. thick. For the 35–6 length, 9 pieces would be required; for the 29–4 side, 8 pieces. The item would be: "3-in precast conc. trench covers, approx. 16 SF each—17 pcs." Here, all labor and materials (formwork, concrete, and setting) would be part of the one item for unit pricing.

Reinforcing steel has not been shown as part of the concrete take-off in any of the examples in this chapter; it is covered in Chap. 4. Where shown in the figures, wire mesh has been included as part of the concrete quantities, because mesh is usually a material purchase item that is set by the general contractor's own crew.

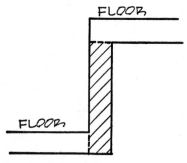

Figure 3.5 Floor drop in pit.

Take-off of concrete retaining wall in W. D. 3.3

Conc. retg. wall

			FORMS	

```
Ftg.            51- 6 × 1-0 × 1-0  =   52 CF              105 SF
                51- 6 × 6-0 × 1-5  =  438     120-0 × 1-5  =  170
Steps        2/ 1- 3 × 6-0 × 1-2  =   18                      14
14½-in. wall   10- 8 × 7-9  =  83 SF
                 7- 4 × 6-6  =  48
                 5- 5 × 7-8  =  42
                14- 0 × 6-5  =  90       2 × 344 SF       =  688
                 3- 7 × 5-9  =  21      + 1-5 × 18-0      =   25
                 3-11 × 6-11 =  27                          1,002 SF
                 5- 7 × 5-11 =  33
                      1-2½ × 344 SF = 416 CF
                                      924 CF
                                  =    34 CY   Ftg. key  =  50 LF
```

Rubbing walls

```
                104-0 × 3-6  =  364 SF
```

Notes on the take-off (W. D. 3.3)

Retaining walls with wide footings should be taken off as one item (footing plus wall). The formwork for a wide footing is more expensive than that for the ordinary 20-in. × 10-in. or 24-in. × 12-in. footings, and in fact the formwork for the footing in the above example would cost as much per square foot as the wall forms themselves; thus those items need not be separated. Notice that the ends are included in the formwork for the footing, and that the steps are also included on both the concrete and formwork sides of that item.

The exposed area for rubbing is the average of the two faces above ground.

CONCRETE COLUMNS

The column schedule (W. D. 3.4) is typical of the schedules found on structural drawings. The totals for each group (the circled figures above the schedule) and the notations (E, I, 8E, 2I, and the like) would not be given on the schedule; they are our own notations made in colored pencil. The totals for each group would be obtained from the plans and inserted on the schedule to simplify the take-off. The E and I notations denote the exterior and interior columns. Note that unless both exterior and interior columns appear in the same group, the I notation need not be used; the exterior groups having been so marked, the rest must be interior. It is necessary to separate the two not only because their heights vary, but also for the following reasons: (1) the exterior columns are probably deductions from the exterior masonry; (2) dovetail slots may be required in the exterior columns; (3) the finishing of the columns probably differs. The column heights should be marked on the schedule if not already shown. When taking off the columns, check off the schedule as each item is counted and entered on the collection sheet, as in Table 3.2. Because of space, the totals of the various columns items have been omitted; using wider spacing, the extensions could be shown as follows:

12×12

$$
\begin{array}{r}
18/\ 10\text{--}10 = 195\text{--}\ 0 \\
18/\ \ 9\text{--}11 = 178\text{--}\ 6 \\
26/\ \ 8\text{--}11 = 231\text{--}10 \\
\hline
605\text{--}\ 4
\end{array}
$$

The columns take-off is now simple.

The take-off (W. D. 3.4)

Conc. cols.

								FORMS
Ext.	0–10	×	1–0	×	161 ft	=	134 CF	590 SF
	1– 0	×	1–0	×	606	=	606	2,424
	1– 0	×	1–2	×	72	=	84	312
	1– 2	×	1–2	×	166	=	226	775
Int.	0–10	×	1–0	×	688	=	573	2,523
	1– 0	×	1–0	×	179	=	179	716
	1– 0	×	1–2	×	388	=	453	1,682
	1– 2	×	1–2	×	194	=	264	905
	1– 4	×	1–4	×	174	=	309	928
							2,828 CF	10,855 SF
						=	105 CY	

Notes on the take-off (W. D. 3.4)

The exterior-column height is the floor-to-floor height less the depth of the spandrel beam; for example, on the third floor the column height is 10–9 less a 1–10 beam, or 8–11. The interior columns are the floor-to-floor height less the 1–4 beam depth.

If there are numerous beams of varying depth, it may be necessary to take them off onto the collection sheet before figuring the columns. From a study

TABLE 3.2 Collection Sheet for Columns

Cols.	10 × 12	12 × 12	12 × 14	14 × 14	16 × 16
Ext.	18/ 8–11	18/ 10–10	8/ 8–11	8/ 10–10	None
		18/ 9–11		8/ 9–11	
		26/ 8–11			
	161 ft	606 ft	72 ft	166 ft	
Int.	43/ 9–5	19/ 9–5	2/ 9–5	16/ 9–5	8/ 10–5
	13/ 10–5		17/ 10–5	2/ 10–5	8/ 11–4
	13/ 11–4		17/ 11–4	2/ 11–4	
	688 ft	179 ft	388 ft	194 ft	174 ft

+ 4 col. base plates 24 in. × 24 in.

of the beam take-off, it should be possible to determine an average beam depth to use as the deduction required to obtain the column height. If a particular beam depth is predominant at the column lines, that depth may be used when the column heights are determined.

If the columns are larger in cross section than the beams, they should be measured to the underside of the slab above, and the beams measured between columns.

CONCRETE BEAMS AND SUSPENDED SLABS

Beam sizes are usually shown in beam schedules on the structural drawings, with the beams numbered—B1, B2, and so on. The easiest way to take off beams is to mark the sizes on the plans, checking off the beam schedule item by item, and then, dispensing with the schedule, compile the collection sheet from the plans. For a large building that has entire structural sheets of structural schedules, it is best to extract the schedule sheets and place the pertinent sheet (folded, if that is more convenient) alongside the plan. This method will simplify the tedious job of marking the beam sizes on the plan, and is much easier than constantly turning back from one drawing to the other.

In the example (W. D. 3.5), the beam sizes have been marked on the plan; note that this is the first-floor plan and that it has been noted: "Second floor similar; roof similar except as noted." The spandrel beams will be taken off at full depth, the slab measured inside the spandrels, and the interior beams taken off to the underside of the slab. The formwork item for beam sides will have to be adjusted if the slab depth changes at a beam. Sometimes such adjustment of the beam sides is very minor and not worth the time spent figuring it, but only experience will enable you to know when to ignore this item and when to measure it. If a particular slab depth predominates, the adjustment of the beam-sides formwork may be ignored. If the design shows a consistent change (such as a 7-in. thickness in the rooms and 5 in. through the corridors), the additional formwork (2 in., for the corridor beams) should be taken off. The collection sheet for W. D. 3.5 is in Table 3.3.

The take-off (W. D. 3.5)

Conc. beams

			FORMS	
			Bottoms	Sides
Spand.	$900\text{-}0 \times 0\text{-}8 \times 1\text{-}10$	= 1,100 CF	600 SF	2,850 SF
Int.	$54\text{-}4 \times 0\text{-}10 \times 0\text{-}8$	= 30	45	72
	$27\text{-}2 \times 0\text{-}10 \times 0\text{-}9\frac{1}{2}$	= 19	23	43
	$122\text{-}4 \times 0\text{-}10 \times 1\text{-}3\frac{1}{2}$	= 132	102	
	$122\text{-}4 \times 1\text{-}0 \times 1\text{-}6$	= 183	122	1,081
	$122\text{-}4 \times 1\text{-}0 \times 1\text{-}7\frac{1}{2}$	= 201	123	
	$27\text{-}0 \times 1\text{-}0 \times 0\text{-}5\frac{1}{2}$	= 12	27	24
	$227\text{-}0 \times 1\text{-}3 \times 0\text{-}11\frac{1}{2}$	= 272	284	435
		1,949 CF	1,326 SF	4,505
				− 231
		= 72 CY		4,274 SF

Conc. susp. slabs

			Slabs = 14,778 SF
6 in.	5,280 SF	= 2,640 CF	Less bm. bot. 726
$4\frac{1}{2}$ in.	5,793	= 2,173	14,052 SF
10 in. + 3 in.	3,705	= 4,014	
	14,778 SF	8,827	*Slab screeds*
			= 1,500 LF
Less pans $1,665\text{-}0 \times 1\text{-}8 \times 0\text{-}10$		= 2,313	
		6,514 CF	20-in. × 10-in. long
		= 241 CY	pans = 1,665 LF

Trowel finish floors = 10,252 SF

Float roofs = 5,126 SF

The collection sheet for W. D. 3.5 (beams and slabs) is shown on p. 71, Table 3.3. In figuring the slabs, the total slab area is taken off first; then, taking off the 6-in. slabs and the 10-in. + 3-in. slabs, their combined area is subtracted from the total area to obtain the $4\frac{1}{2}$-in. slab area. This method is

TABLE 3.3 Collection Sheet for W. D. 3.5

Span. Bms.				Int. Bms.				Beam sides (adjustment) Add			Deduct	
8 × 22	10 × 8	10 × 9½	12 × 18	12 × 19½	10 × 15½	15 × 11½	12 × 5½	1½ in.	4½ in.	8½ in.	1½ in.	7 in.
88– 9	27–2	27–2	61–2	61–2	122–4	37–10	27–0	61–2	36	25–5	24–3	47–6
61– 2	27–2		61–2	61–2		× 6		61–2	× 3	37–10	24–3	26–0
149–11	54–4		122–4	122–4		227–0		122–4	108	63–3	48–6	9–8
149–11										× 3		83–2
299–10										189–9		× 3
× 3												249–6
899–6												

```
          +          −
         16         135
         40           6
         56         146
                    287
                    −56
To take-off sheet —  231 SF (deduct)
```

TABLE 3.3 Collection Sheet for W. D. 3.5 *(Continued)*

Total slabs

1st	77–9 × 61–2	=	4,756 SF			
	26–0 × 9–8	=	251			
			5,007			
Less Stair	9 ft × 9 ft	=	81			
			4,926			
2nd		=	4,926			
3rd (roof)		=	4,926			
			14,778 SF			

10 in. + 3 in.

47–6 × 26–0 = 1,235 SF
× 3
3,705 SF

6 in.

61– 2 × 28–2 = 1,723 SF
37–10 × 24–3 = 917
1st 2,640
2nd 2,640
5,280 SF

4½ in.

total = 14,778 SF
6 in. = 5,280
10 in. + 3 in. = 3,705
8,985
5,793 SF

Pans

12 × 46–3 = 555 ft
× 3
1,665 ft

invariably both quickest and best; first take off the total area of all the suspended slabs, then measure the simpler, less numerous areas, and finally subtract the sum of these from the total area to obtain the area of the predominant type of slab.

Notes on the take-off (W. D. 3.5)

The total length of the spandrel beams being 899–6, the item is entered at 900–0 on the take-off sheet (p. 70); it is reasonable to round out this item, as 6 in. in 900 ft is of little consequence.

The beam bottoms are figured first and then used to obtain the concrete quantities (beam-bottom area times depth). The spandrel-beam sides are 1–10 on the outside and 1–4 on the inside; 900–0 × 3–2 gives 2,850 SF. The interior-beam formwork is the length times twice the depth (for two sides). The three beam items that are 122–4 long are combined for convenience in computing the beam sides; adding the beam depths together (1–3½ plus 1–6 plus 1–7½, totaling 4–5) gives the area of the forms for the beam sides as 122–4 times twice 4–5, or 1,081 SF.

The deduction from the beam sides (231 SF) was computed in detail on the collection sheet.

The slab forms are the total slab area less that of the interior beam bottoms. If the slab is taken off to the outside face of the spandrels, then the total of all beam bottoms is deducted to obtain the slab forms, the spandrels are measured to the underside of the slab (8 in. × 16 in. in the example), and 6 in. is added to the face side of the spandrel-beam formwork (22 in. plus 16 in., or 3 ft 2 in.).

The concrete for the 10-in.+3-in. slab (10-in. pan with 3-in. concrete over) is first figured as a solid 13-in. slab, and then the volume of the pans is deducted. Very often a floor plan does not show all the pan layout, but simply a typical layout of one small area. If that is so, there are two methods that may be used to compute the items for the concrete and the pans: (1) the pans can be laid out by marking them off, either throughout the entire floor plan or in complete areas that are repeated, or (2) a typical area can be computed so as to arrive at factors that can then be projected for the entire pan area. To illustrate method 2, consider the quantities for the pan slabs in W. D. 3.5 as being for a typical area of a large floor, a floor having, say, 20,000 SF of that type of slab (see example on p. 74).

This second method, although much simpler than laying out all the pans, is not as accurate as the first method, and it is not recommended except for rough-checking of quantities. Slab layouts usually repeat design patterns, so the entire pan quantity can often be computed by laying out the pans in one of each of the different areas and making the repetitions work for you.

Typical area		3,705 SF	×	1– 1	=	4,014 CF
Less pans	1,665–0	× 1–8	×	0–10	=	2,313
						1,701 CF

    concrete = 0.46 CF per SF of floor
    pans   = 0.75 SF per SF of floor
or
    pans   = 0.45 LF per SF of floor

For the entire pan slab area of 20,000 SF

Concrete	20,000 SF	×	0.46	=	9,200 CF
20-in. pans	20,000 SF	×	0.45	=	9,000 LF

Regarding the total slab area, the collection sheet should separate floors from roofs. Also note that the slab finish area is the slab area plus the spandrel-beam-bottom area (because the slabs were measured to the inside of the spandrels).

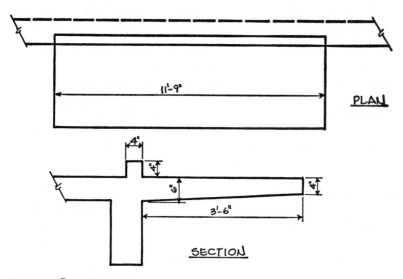

Figure 3.6 Concrete canopy.

Concrete canopy

The spandrel beam would have been taken off with the building slab and beam items. The canopy (see Fig. 3.6), although to be poured with the slabs, is kept separate because the formwork is expensive. The take-off would read:

Concrete

Conc. canopy

11–9 × 3–6 × 0–5 = 17 CF	*FORMS*
+11–9 × 0–4 × 0–4 = 1	Soffit 41 SF
18 CF	Edges 9
= 1 CY	Curb 8
	58 SF (cantilevered)

Rubbing canopy = 58 SF

Notes on the take-off

The small curb over the spandrel is part of the canopy (see Fig. 3.6) and is taken off with the canopy. The thickness is averaged for the concrete item, but the forms are taken off at the maximum thickness (6 in.) around the three sides. The rubbing item is the same area as the formwork—the exposed soffit and sides.

CONCRETE INTERIOR STAIRS

Notice that this is stair No. 1 (see Fig. 3.7). Stair No. 2 is similar; stair No. 3 is also similar, except that it is 6 in. wider. The wall under the second flight

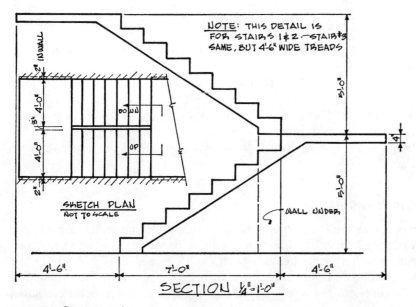

Figure 3.7 Concrete stairs.

may be assumed to have been taken off previously. The section is drawn to a scale of ¼ in. to 1 ft; the underside of the stairs is exposed and must be rubbed. The take-off would read:

Conc. int. stairs

                                                                         FORMS

Nos. 1 & 2	2/2/8–8 × 4–0 × 1–0 = 139 CF	4 × 35 SF			
		4 × 9	}	256 SF	
		4 × 20			
	2/2/8–7 × 4–6 × 0–4 = 52	4 × 37		= 148	
No. 3	2/8–8 × 4–6 × 1–0 = 78	2 × 39			
		2 × 9	}	142	
		2 × 23			
	2/9–7 × 4–6 × 0–4 = 29	2 × 42		= 84	
	298 CF			630 SF	
	= 11 CY				

Rub risers = 126 SF

Trowel treads and landings
        2/16–0 × 8–3 } 412 SF
           16–0 × 9–3

Rub stairs, soffits, etc. = 504 SF

Notes on the take-off

The stair soffit (see Fig. 3.7) is scaled off the drawing at 8 ft 8 in., and—as for basement stairs—the thickness is taken off at 12 in. to allow for spillage and waste. The stair forms are the soffit, (8–8 × 4–0) 35 SF, the open side, (8–8 × 1–0) 9 SF, and the risers, (4–0 × 5–0) 20 SF—all multiplied by 4 (for 4 flights, 2 flights each for stairs Nos. 1 and 2). The landings are taken off full size for concrete but "wall to wall" for forms.

The rubbing item is the formwork total area less that of the risers (630 SF less 126 SF, or 504 SF).

MISCELLANEOUS FORMWORK ITEMS

Grade beams are taken off in the same way as foundation walls, except that the formwork may be a little different. It will be necessary to study the drawings to decide whether grade-beam bottoms will be needed. If in doubt at the time of taking off, keep the beam-bottom forms item separate from the sides. The item "Grade-beam bottoms" may then be entered on the estimate sheet with a query or similar notation that will draw special attention to it.

In taking off grade-beam sides, no special procedure is involved if the beams are to be poured separately from the slab. For grade beams poured with the ground slab, however, the area of the inside forms, which will be lost, should be listed separately from the outside form, which will be recovered. Forms that will be covered up and therefore cannot be recovered should always be shown separately.

For framed ground slabs, beam bottoms will not usually be required. The lost beam sides should be taken off separately, the same as for grade-beam forms.

Formwork for high slabs should be a separate item from that for the normal slab areas. "Normal slab areas" may be considered as being slabs that can be formed using a single jack system. Whether wooden tee jacks or tubular metal shores are used, 13 ft to 14 ft is the maximum height obtainable with a single jack. Slabs requiring a shore of over 14 ft must be taken off separately (for formwork at least) and the height noted.

Foundation-wall formwork should be separated into two items—walls up to 10 ft high, and those over 10 ft high—so that the labor costs can be properly considered. Walls requiring difficult or intricate formwork—such as curved walls, Y-walls, or walls having fluted or other architectural finishes—should always be taken off separately and the formwork described.

Wall openings should be deducted from the concrete and the extra formwork items taken off. There is no deduction from the normal formwork item; instead, a separate extra-cost item is taken off. That item can be the measured perimeter of all the openings, taken off as "Boxing for wall openings—...LF"; or the openings may be counted in groups, and the item taken off as "Box out for wall openings up to 10 SF—...opgs.," then similarly for openings from 10 SF to 25 SF, and so on.

Brick shelf is usually taken off in square feet (see p. 59). A one- or two-course brick shelf, however, should be taken off in lineal feet and the size described. A brick shelf of up to two courses will not normally require special formwork make-up.

Haunches on walls are described by size and taken off in lineal feet—for example, "5-in. × 8-in. haunch to wall—120 LF." If necessary a sketch may be drawn in the left-hand margin of the estimate sheet. Haunches of odd size and design can be very expensive formwork items. All items such as haunches, pilasters, and brick shelves that are extra-cost items for formwork are included in the main item on the concrete side; that is, they are added into or deducted from the wall concrete item. These items will be ultimately contained within the one final item "Concrete foundation walls"—but they appear as separate extra-cost formwork items in the estimate.

Pilasters are added into the wall concrete item. The pilaster forms may be taken off as an extra-cost item in square feet and the number of pilasters noted: "Extra for wall pilasters (14)—250 SF."

Slab bulkheads for concrete pourings are taken off in lineal feet. If the slabs have to be poured checkerboard fashion, the quantity of bulkheads can be measured. If the pouring is not required to be done in any particular pattern, the item must be figured by estimating the extent of each day's pour and allowing bulkheads accordingly. If reinforcing is to be continuous through the bulkheads, it should be noted and described in the slab bulkhead item.

Slab bulkheads at openings (such as holes through the slab where the formwork is not stopped) are measured as the perimeter of the openings in lineal feet. Usually, openings for stairs and elevator shafts are not decked in, but minor openings would be decked in and bulkheaded.

Chamfer strips for columns, beams, and the like are taken off in lineal feet.

Screeds for pitched slabs are measured in lineal feet and described as "Slab screeds to pitch." Such an item is common for roof fill pitched to roof drains, and also for floors pitched to drains; it should be kept separate from the ordinary slab screeds item.

Metal forms "left in place" for slabs are corrugated metal forms supplied by the manufacturer in widths suitable for spans up to about 8 ft. The item should be taken off in square feet allowing 8 to 10 percent for waste. Some types of corrugated metal slab forms will require temporary shoring, which must also be taken off; such shoring, however, may only amount to a row of jacks with a 3-in. × 4-in. or 4-in. × 6-in. header. Having decided what is required, take off an item—"Temporary shoring for metal slab forms"—in lineal feet, with all rows measured.

Gallery step seating formwork must be carefully taken off. Take it off similar to ordinary stairs, and fully describe each item. Curved seating should be taken off separately and also described. Aisle steps may be taken off separately by number: "Formwork for aisle steps 2–9 × 1–0 × 0–6—74 each."

The concrete for straight and curved seating may be taken off in a single item; the small aisle steps, however, if poured separately, should be taken off separately.

MISCELLANEOUS CONCRETE ITEMS

Roof fill is taken off in cubic yards. Care must be taken in determining the average depth of pitched roof fill. If the pitch is one way only (as from front to back or from both front and back to a parallel low line through the center), then the average depth is simply the average of the two thicknesses. In Fig. 3.8, for example, roof fill is 6 in. at the high point, 2 in. at the low point; the average thickness of fill for either A or B is 4 in.

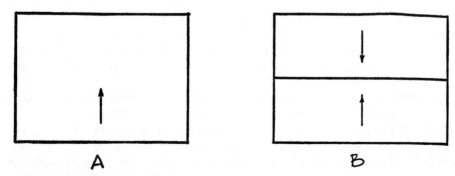

Figure 3.8 Roof fill. One-way pitch.

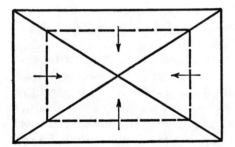

Figure 3.9 Roof fill. Four-way pitch.

Now consider Fig. 3.9, in which the roof fill pitches four ways, from 6 in. at the outside to 2 in. at the center. The average thickness will vary according to the roof dimensions. There is no area that is 2 in. thick—only the point at the center is 2 in. thick. The average thickness of the fill will be the thickness along the sides of the rectangle (dotted in Fig. 3.9) that encloses half of the roof area. The average thickness of roof fill for Fig. 3.9 would be $4\frac{3}{4}$ in. (The figure may be checked by laying out a sample roof.) For estimating purposes, the table shown will serve.

Roof fill pitched as in Fig. 3.9

Pitch	Fill — average thickness
2 in.	Minimum + $1\frac{1}{2}$ in.
3	Minimum + 2
4	Minimum + $2\frac{3}{4}$
5	Minimum + $3\frac{1}{2}$
6	Minimum + $4\frac{1}{4}$
7	Minimum + 5

Note that from roof B in Fig. 3.8 to the roof in Fig. 3.9 the fill increases from 4 to $4\frac{3}{4}$ in.; fill pitched between those two extremes would vary accordingly, as shown in Fig. 3.10.

Lightweight concrete must be thoroughly investigated for yield. Some of the expanded vermiculite or foam-type lightweight concretes require a 10 percent addition to the net volume.

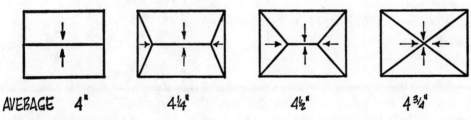

Figure 3.10 Roof fill. Varied pitches.

For insulating concrete around underfloor ducts, measure the gross quantity of concrete and deduct the volume of the ducts. Side forms will probably be necessary; if so, take them off in square feet.

Concrete walks are taken off item by item: gravel bed in cubic yards; edge forms in lineal feet (curved forms separate from straight); concrete in cubic yards; finishing (float or trowel) in square feet; expansion jointing in lineal feet. If the walks are taken off with the site work, excavation for them could be taken off at the same time. It is generally best to include all site work items in the Excavation and Site Work part of the estimate. Both for cost recording and for comparison to previous estimates, it is best to have the items for the building proper segregated from the outside work.

Abrasive aggregate is taken off in pounds and fully described, or taken off in square feet with the quantity of abrasive per square foot of floor as specified.

Machine pads are taken off in square feet and the thickness stated.

Floor hardener is taken off in square feet and described; for example: "Magnesium fluosilicate floor hardener (2 coats)—...SF."

Cement fill for metal pan stairs is taken off item by item: "Cement fill (pan stairs) 1:1—...CY; Mesh for stair fill—...SF; Trowel finish stair fill—...SF; Set metal nosings (0-0 long)—...pcs."

Chimney caps (concrete) are best taken off as an item for lump-sum pricing of labor and material, although they could be taken off in detail. A building usually has only one chimney cap (or two at most), so that a lump-sum price is satisfactory.

Machine foundations may be taken off in the same way as concrete footings, but should be figured separately from the footings. Particularly if they are heavy foundations with especially heavy formwork and large quantities of concrete. The formwork will be more expensive and the concrete will probably cost less to place than for ordinary footings.

Heavy slabs, floors over 12 in. thick, should be separated from slabs up to 12 in. thick, because of the lower labor cost of placing the concrete and the higher cost of screeds and bulkheads.

Curbs may be taken off in lineal feet and described; or they may be measured in detail and described, with the formwork taken off as a separate item and the finishing measured. In the case of exterior site curbs, which would be taken off with the site work, the excavation item would be taken off then, too. The curb items would be kept together, and on the estimate they would read:

Curbs—excavation	CY
Curbs—concrete (6 in. × 17 in.)	CY
Curbs—forms	SF
Curbs—rubbing	SF

After having priced a few jobs that include this item, you may be able to set up a lineal-foot price, and thereafter take the item off as "6-in. × 17-in. conc. curb (include exc., forms, etc.)—...LF."

Oil-tank mats are taken off the same as footings; concrete in cubic yards, forms in square feet. An additional item for "Setting oil-tank anchors" should be taken off and described if necessary.

Oil-tank manholes and oil-pipe trench (if concrete) may be taken off as one item. It is not necessary to make separate items of the bottom slab, sides, and top slab. Measure the various parts properly, but add them all into one total: "Oil-tank manholes and pipe trench, forms—...SF." Do the same for the concrete.

For concrete encasing ducts (electric or telephone), take off concrete in cubic yards (with duct volume deducted) and forms in square feet.

Hi-early cement for speeding up the setting of concrete must be allowed for in the estimate if it is to be used. Take off as an extra-cost item: "Extra for Hi-early cement—...CY."

Heated aggregate for cold-weather concrete probably will not cost extra, but this must be checked with the ready-mix suppliers. If a charge is made, an extra-cost item should be carried for the anticipated quantity of concrete involved: "Extra for heated aggregate to concrete—...CY."

For locker bases and all similar bases: take off concrete in cubic yards and edge forms in either lineal feet (if up to 12 in. high) or square feet (if over 12 in. high).

Concrete curing is taken off in square feet, and described if the curing method is specified. If burlap cover is called for, take it off, adding 10 percent for laps and waste, and enter it as "Burlap, curing slabs—...SF"; if taped joints are called for, the item should be "Burlap (taped joints), curing slabs—...SF."

Granolithic floor topping is taken off in a series of items: "Grano. floor topping (1 in.) 1:1:2—...CY; Screed for 1-in. topping—...LF; Trowel finish grano. floors—...SF."

Color admixtures for floors are taken off in pounds or gallons, depending on the type of color specified.

Cement base is taken off by sizes in lineal feet, and must be fully described; for example: "1 × 6 cement base, bullnosed (incl. mesh)—...LF." Sometimes a small sketch of the base (in the left-hand margin of the estimate sheet) will prove helpful when pricing the item.

Waterproofing admixture for concrete may be taken off in pounds or gallons, or for the quantity of concrete to be waterproofed but always as an extra-cost item. Usually only particular areas are specified as requiring concrete to be waterproofed such as foundation walls, certain ground slabs, or perhaps pits. Do not separate the concrete items. Take off the concrete as usual (note the waterproofed items on the takeoff sheet) and then take off a single extra-cost item. "Extra for concrete waterproof admixture ...CY."—and describe the type of waterproofing specified. If the quantity of admixture needed per cubic yard is known, the item may be expressed in pounds or gallons rather than in cubic yards.

Concrete tests are usually a lump-sum item on the estimate, so the take-off sheet may read simply "Concrete tests—L. S." If the specifications call for control of the concrete—with a testing engineer at the plant or at the site, or both—at the contractor's expense, and a price is quoted per cubic yard of concrete; then the item may be taken off as: "Testing and control of concrete—...CY." If the requirements are more complex, such as costs per test cylinder the item should be marked with a reference to the applicable page in the specifications.

Precast concrete cannot be taken off to one set system; the method that is best for one item may not be suitable for other precast items. Simple precast lintels may be taken off by lengths, grouping the items together—3 ft to 4 ft, 4 ft to 5 ft, and so on—under the average length of each group. The estimate would show "Precast concrete lintels (0–4 × 0–8 × 3–6 long—...pcs."; the following item might read "Ditto × 4–6 long—...pcs."; and so on.

Precast concrete columns, beams, and copings must be taken off in such a way as to enable the items to be intelligently priced. As an example, consider precast concrete columns varying in both cross section and height. The concrete side of the take-off would list the items separately; they would each be measured and totaled, and the concrete quantity listed in cubic yards. The formwork side would show a cross-section sketch, the maximum length, and the number of columns of each type; there would be one formwork item for each type of column. The hardware (base plates, bolts, and the like) would be taken off and listed separately. An item for the erection would also be taken off, and could be carried as so many precast concrete columns at an average weight. As completed and transferred to the estimate, the precast concrete columns item might read as shown on p. 83. The general idea, is to make it possible to properly visualize the requirements of each item when pricing it.

Many precast concrete items will be taken off by cast-stone companies, and bid on as a delivered sub item. In that case, the general contractor's estimate would include only the items for unloading and setting if not included by the precast sub. Especially heavy precast items or those that are to be set high above ground may require a crane item: "Crane setting precast items—...days." Similarly, if precast items will need considerable rigging or temporary guying, it is be advisable to include an additional item for that work. "Box up and protect precast concrete" is another item that should sometimes be taken off. For finished sills, steps, and the like, which are likely to be chipped after being set, a "Box up and protect" item is always necessary.

SUNDRY ITEMS TAKEN OFF WITH CONCRETE

Visqueen or polyethylene under ground slabs is taken off in square feet with 8 to 10 percent added for laps and waste.

Membrane or special toppings over slabs is also taken off in square feet with 8 to 10 percent for laps and waste. If the membrane is to be mopped on, that fact should be noted. The same applies to any special requirements for felt, tar paper, and the like.

Perimeter insulation for foundation walls is measured in square feet. The waste may vary considerably, and will depend on the type of insulation and its depth. Usually, however, a 10 percent waste allowance is sufficient. Some of the insulation board materials, however, come in standard 4-ft widths only, and cutting them to size may leave considerable waste. For instance, if 3 ft of insulation board is called for and "piecing up" is not allowed, there will be 25 percent waste. Often, even if it is permissible to piece up the insulation, it would not be economical to do so, because of the labor involved. The waste must be considered in conjunction with the labor factor to determine whether it would be cheaper to piece up the material or to lose the waste and save labor by using only full-width material.

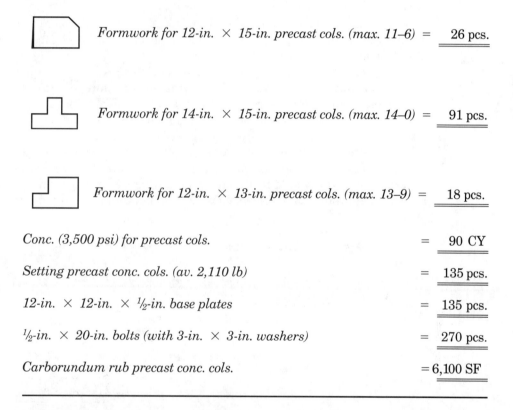

Formwork for 12-in. × 15-in. precast cols. (max. 11–6) =	26 pcs.
Formwork for 14-in. × 15-in. precast cols. (max. 14–0) =	91 pcs.
Formwork for 12-in. × 13-in. precast cols. (max. 13–9) =	18 pcs.
Conc. (3,500 psi) for precast cols. =	90 CY
Setting precast conc. cols. (av. 2,110 lb) =	135 pcs.
12-in. × 12-in. × ½-in. base plates =	135 pcs.
½-in. × 20-in. bolts (with 3-in. × 3-in. washers) =	270 pcs.
Carborundum rub precast conc. cols. =	6,100 SF

Waterstops are taken off in lineal feet by types and sizes. Copper waterstops are taken off in square feet and the weight stated; for example, "20-oz. copper waterstops—...SF".

Expansion jointing is taken off either in square feet with the thickness stated or in lineal feet with both the width and thickness stated. Some expansion material comes only ½ in. thick, so that for 1-in. expansion jointing the length or the area must be doubled and shown as ½ in. thick on the estimate sheet. Other expansion material comes in thicknesses up to 1 in.; the standard widths also vary. The method of taking off should properly describe the material specified. Liquid poured sealer or caulking for expansion joints is taken off in lineal feet.

METAL SUNDRIES FOR CONCRETE WORK

Dovetailed anchor slots and anchors are for securing masonry to concrete are taken off in lineal feet and thousands of pieces, respectively. Usually both the spacing and the required locations of dovetail slots are fully specified. If the spacing is not specified, allow one vertical dovetail anchor slot (that is, one continuous anchor slot) for every column face that has masonry veneer. For spandrel beams, brick shelves, and the like, allow for the vertical placing of anchor slots full height on 2-ft centers. Unless spacing is specified, allow one anchor for every lineal foot of slot required.

Wedge inserts for securing steel lintels and the like to concrete are taken off in pieces of stated size. If spacing is not specified, allow 1 insert per 3 LF, plus "one for the end" for each lintel.

Ceiling inserts should be specified if required. Take them off in pieces, always allowing "one for the end." In laying out or taking off any item that is spaced out center-to-center, an extra piece will always be needed "for the end." Inserts on 4-ft centers on a ceiling 12 ft × 8 ft would require 4 in one direction and 3 in the other—total of 12 pieces.

Abrasive metal stair nosings are taken off by size in pieces, and described.

Light iron items to be supplied by a subcontractor but set in concrete by the general contractor should be taken off for setting. If not supplied by a subcontractor, they would have to be taken off item by item and fully described. For example:

4-in. × *4-in.* × *¼-in. angle-iron column guards 6–0 high* = 22 pcs.

4-in. × *4-in.* × *¼-in. angle-iron column guards 7–4 high* = 18 pcs.

½-in. × *16-in. anchor bolts for cornice* = 200 pcs.

Pipe sleeves 2-in. dia. × *12 in. for fence posts* = 132 pcs.

Set only 14-in. × *14-in. col. base plate* + *1 pr. anchor bolts* = 81 cols.

Set only 2-in. × *2-in. angle-iron floor frame* = 130 LF

Refer to the structural-steel and miscellaneous-iron specifications to check which items are simply set by the general contractor and which are not included in the sublet sections.

Subcontract items generally must be checked to determine whether they are to be embedded in or secured to the concrete, in order to ensure that all the work required in the concrete section of the specifications is taken off. Discrepancies between the subcontractors' sections of the specifications and the concrete section must be clarified by the architect, or, if this cannot be done, agreed upon with the subbidder. If the item in question cannot be satisfactorily cleared, either it must be taken off or the subbid adjusted to include it.

Chapter 4

Reinforcing Steel

Taking off reinforcing steel may be either simple or complicated, depending on the nature of the job and the purpose of the take-off. A general contractor will generally take off reinforcing steel to obtain the tonnage if no subbids have been received for the item. Even if prices for reinforcing steel have been received, the general contractor may take off the item in order to have a check or control figure on the weight and price quoted by the supplier. If reinforcing steel is a major item in a bid, every effort should be made to ensure that subbids will be received such as sending cards to prospective suppliers or with follow-up telephone calls.

In some districts labor agreements require reinforcing steel to be bent on the job rather than in the shop. The bending thus becomes part of the field setting, and must be considered even if the take-off is for setting only. Generally, the steel supplier's bid on reinforcing steel will include shop drawings, schedules, bending, and delivery. If shop bending is not permitted, then the bid will include everything except the bending, and the bars will be delivered in straight lengths (cut to size and tagged) for bending in the field. Steel suppliers' bids must be carefully checked, especially if the bid is on a unit-price basis. Sometimes the bid is given on a lump-sum basis, but more often it is quoted per 100 lb, with the estimated total weight given. The subcontractor's bid will define the terms and conditions of the quotation.

Also important is any escalator clause in a steel bid. It is common for steel suppliers to bid with a provision that the prices quoted are subject to increase should the mill price of steel increase after acceptance of the bid. Accessories must also be considered in evaluating reinforcing-steel bids. If the accessories are not included by the supplier, the general contractor must either take them off or make an allowance for them.

Reinforcing steel is taken off in lineal feet by sizes and then converted into weight—either pounds or tons (2000 lb). A take-off for supply and fabrication should include considerable detail. Separate totals are required for Nos. 2, 3, 4, and 5 bars; Nos. 6 to 11 may be lumped together. Base prices for reinforcing steel are adjusted by "size extra" costs—that is, the extra costs for sizes 2,

3, 4, and 5. The take-off must further designate steel bending by two categories: (1) light bending, and (2) heavy bending. Although it sounds contradictory, light bending is the more costly. It includes all stirrups and column ties (regardless of size), all bending for Nos. 2 and 3 bars, all bars bent at more than six places, all bars bent in more than one plane, and all bars with more than one radius bend (that is, a bend of 10-in. radius or greater). All other bending is classified as heavy bending.

Reinforcing steel is normally measured in 30-ft lengths, with allowance for laps. Laps are allowed for by adding 40 diameters of the bar for every 30 ft or fraction of 30 ft in length. Although the latest code permits laps of 30 diameters, many estimators continue to allow 40. In taking off top steel for slabs that are to have "one-way" reinforcing, the estimator should take off chair bars even if they are not shown on the drawings. High chairs are spaced at about 4-ft centers alongside the beams or at 30-ft intervals in the other direction, or both; a No. 5 bar is placed on each line of high chairs, and the top steel rests on these chair bars; otherwise, hundreds of high chairs would be needed. If temperature steel should run across the main steel, or the top steel should run in both directions, then chair bars would not be needed.

Steel bars

Bar size		Weight, in lb per LF	Lap (of 40 dia.), in ft
Old, size in inches	New, by number		
$\frac{1}{4}\phi$	2*	0.167	0–10
$\frac{3}{8}\phi$	3	0.376	1– 3
$\frac{1}{2}\phi$	4	0.668	1– 8
$\frac{5}{8}\phi$	5	1.043	2– 1
$\frac{3}{4}\phi$	6	1.502	2– 6
$\frac{7}{8}\phi$	7	2.044	2–11
1 ϕ	8	2.670	3– 4
1 □	9†	3.400	3–10
$1\frac{1}{8}$ □	10†	4.303	4– 3
$1\frac{1}{4}$ □	11†	5.313	4– 9

ϕ — round; □ — square.

* Plain round only (not "deformed").
† Round bars equivalent in weight to the 1-in., $1\frac{1}{8}$-in., and $1\frac{1}{4}$-in. square bars.

Such factors as waste and cutting do not concern the estimator in taking off reinforcing steel, except in special cases. Normally, no waste is allowed if there is a good set of drawings and specifications, and if a careful take-off has been made with allowances for laps. If the drawings are poor, however, and there is some doubt about the reinforcing-steel requirements, it might be prudent to add from 2 to 4 percent to the total quantity.

If, as in the examples that follow, the reinforcing steel is to be taken off for setting, including job bending, then the final totals should separate the items to facilitate intelligent pricing. For this purpose, it is unnecessary to separate bars Nos. 2, 3, 4, and 5 (as is done in taking off for supply). In taking off reinforcing steel for setting, bars up to No. 4 may be lumped together, and bars Nos. 5 to 11 collected also. Bent bars should be separated into two items—light bending and heavy bending.

Any special reinforcing-steel items must be considered separately. This classification would include such items as column spirals, prefabricated shear heads, welded steel, special dowels (greased and wrapped), and epoxy-coated reinforcing.

Wire mesh is sometimes taken off with the reinforcing steel, but it is simpler and more practical for the general contractor to take it off at the same time as the concrete items that it will be used for.

A table of steel bar data is shown on p. 86.

TAKE-OFF FOR FOUNDATION (W. D. 3.1)

REINFORCING STEEL						No. 3	No. 4	No. 5	No. 6	No. 8
Ftgs.	A	No. 6	7 ×	8 ×	3–0				168 ft	
	B	No. 6	3 ×	4 ×	3–0				36	
		No. 6	3 ×	4 ×	4–6				54	
	C	No. 8	2 ×	6 ×	4–0					48 ft
		No. 8	2 ×	5 ×	5–0					50
Piers		No. 6		4 ×	74–0				296	
(L. B.)*		No. 3	2 ×	12 ×	4–0	96 ft				
Wall ftgs.		No. 5		2 ×	325–0			650 ft		
Walls		No. 4		220 ×	7–0		1,540 ft			
		No. 4		220 ×	7–6		1,650			
		No. 4	2 ×	7 ×	325–0		4,550			
						96 ft	7,740 ft	650 ft	554 ft	98 ft
						×	×	×	×	×
						0.376	0.668	1.043	1.502	2.67
					(LB)	36 lb	5,170 lb	680 lb	832 lb	262 lb

Straight steel (up to No. 4) ≈ 5,170 lb

Straight steel (Nos. 5, 6, & 8) = 1,774 lb

Light bending (No. 3) = 36 lb

* Light bending

88 Chapter Four

Notes on the take-off (W. D. 3.1)

The pier footing items are taken off at full length (although the actual bars as specified or called for on the drawings would be about 4 in. shorter). The pier height is obtained from the concrete take-off (74 LF). The No. 3 stirrups are notated "L. B." to signify light bending. The wall footings and the wall horizontal steel are both 315 LF net (see concrete take-off). Referring to W. D. 3.1, we find that 79–5 requires two laps, 60–4 two laps, 54–1 one lap, and 41–8 one lap, for a total of six laps at 1–8, or 10 LF. Thus, 315 LF plus 10 LF gives 325 LF total length for reinforcing steel.

The vertical wall steel quantities are obtained from the concrete sheet. There is a total of 2,232 SF of wall for the 314–10 perimeter of wall averaging 7 ft in height. For a perimeter of 315 LF with bars on 18-in. centers, 210 bars would be required; adding 1 for the end of each piece of wall, 210 plus 10 is 220 pieces total (for each face). The outside bars will be in two pieces with a 6-in. lap, so we use 7–6 for the height of that face; that is, 220 pieces at 7–0 for the inside face, and 220 at 7–6 for the outside face.

TAKE-OFF FOR RETAINING WALL (W. D. 3.3)

Reinforcing steel (retaining wall)

							No. 3	No. 4	No. 5
Ftg.	No. 4			9	×	54– 0		486 ft	
Ftg.	No. 3			52	×	4– 0	208 ft		
Ftg.	No. 5 (H. B.)*			52	×	6– 6			338 ft (H. B.)
Wall	No. 3	2	×	7	×	51– 9	725		
Wall	No. 4			42	×	6–10		287	
Wall	No. 4			42	×	9– 3		388	
							933 ft	1,161 ft	338 ft
							×	×	×
							0.376	0.668	1.043
							350 lb	775 lb	353 lb

Straight bars (Nos. 3 & 4) = 1,125 lb

Heavy bending (No. 5) = 355 lb

*Heavy bending

Notes on the take-off (W. D. 3.3)

The footing steel item is the measured total length of the footing plus the two 15-in. laps at the steps. The short steel is taken off at the wall length (50–6), which is 51 pieces (on 12-in. centers) plus 1 for the end. The wall height averages 6–10 (see concrete take-off, which shows 345 SF of wall area for 50–6 in length, or an average height of 6–10). The long vertical steel is 6–10 plus 1–5 plus 1–0, or 9–3.

TAKE-OFF FOR COLUMNS (W. D. 3.4)

		REINFORCING STEEL					*No. 8*	*No. 6*	*No. 3* (L. B.)
Cols.	No. 8	10	×	4	×	11– 4	454 ft		
	No. 8	8	×	4	×	10–10	680		
	No. 8	8	×	4	×	10– 5			
	No. 6	18	×	4	×	10–10		780 ft	
	No. 6	26	×	4	×	9–11		1,032	
	No. 6	52	×	4	×	8–11		1,855	
	No. 6	30	×	4	×	11– 4		1,360	
	No. 6	32	×	4	×	10– 5		1,334	
	No. 6	80	×	4	×	9– 5		3,014	
Dowels	No. 6	264	×	4	×	3– 6		3,696	
							1,134 ft	13,071 ft	

L. B. ties No. 3 61 × 12 × 3–8 = 732
 13 × 13 × 3–8 = 169 } 1,083 × 3–8 = 3,971 ft
 13 × 14 × 3–8 = 182

 45 × 12 × 4–0 = 540
 18 × 13 × 4–0 = 234 } 1,026 × 4–0   4,104
 18 × 14 × 4–0 = 252

 10 × 12 × 4–4 = 120
 17 × 13 × 4–4 = 221 } 579 × 4–4   2,509
 17 × 14 × 4–4 = 238

 16 × 12 × 4–8 = 192
 10 × 13 × 4–8 = 130 } 462 × 4–8   2,156
 10 × 14 × 4–8 = 140

 8 × 13 × 5–4 = 104 } 216 × 5–4   1,152
 8 × 14 × 5–4 = 112

 13,892 ft

Straight bars (*Nos. 6 & 8*)

 1,134 ft × 2.67 = 3,028 lb
 13,071 ft × 1.502 = 19,633
  = 22,661 lb
  = 22,670 lb

Light-bent bars (*No. 3*)

 13,892 ft × 0.376 = 5,223 lb
  = 5,230 lb

TAKE-OFF OF STEEL FOR SUSPENDED SLAB (W. D. 3.5)

	REINFORCING STEEL (susp. slabs & beams)						No. 3 L.B.	No. 4 L.B.	No. 6	No. 7	No. 8	No. 9	No. 10	No. 6 H.B.	No. 2
Span. bms.	No. 8	3	×	4	×	324–0					3,888 ft				
	No. 4 (L.B.)	3	×	456	×	5–0		6,840 ft							
Int. bms.	No. 8	3	×	4	×	27–2					326 ft				
	No. 3 (L.B.)	3	×	37	×	4–0	444 ft								
	No. 8	6	×	4	×	68–0					1,632				
	No. 3 (L.B.)	4	×	83	×	6–0	1,992								
	No. 3 (L.B.)	2	×	83	×	5–0	830								
	No. 10	6	×	2	×	42–0							504 ft		
	No. 9	6	×	2	×	41–8						500 ft			
	No. 4 (L.B.)	6	×	57	×	5–2		1,767							
	No. 7	3	×	4	×	9–0				108					
	No. 3 (L.B.)	3	×	12	×	3–8	132								
5-in. rib joists	No. 6	3	×	2	×	26–0			156 ft						
	No. 7	3	×	26	×	24–0				1,872					
	No. 6 (H.B.)	3	×	26	×	30–8								2,392 ft	
	No. 6	3	×	13	×	7–0			273						
							No. 3								
	No. 3	3	×	38	×	26–6	3,021 ft								
	No. 2	3	×	10	×	47–6									1,425 ft
								No. 4							
Flat slabs	No. 4	3	×	124	×	38–6		14,322 ft							
	No. 4	3	×	63	×	38–6		7,277							
	No. 4	3	×	32	×	64–6		6,192							
	No. 4	3	×	57	×	8–9		2,258							
	No. 4	3	×	29	×	8–9									
	No. 4	3	×	77	×	24–0		8,352							
	No. 4	3	×	39	×	24–0									
	No. 4	3	×	19	×	39–6		2,252							
	No. 4	3	×	8	×	27–8		664							
						Bent	3,398	8,607						2,392 ft	
						Straight	3,021	41,317	429	1,980	5,846	500	504		1,425 ft

Straight bars (Nos. 2 to 4)

No. 2	1,425 ft	×	0.167	=	238 lb
No. 3	3,021	×	0.376	=	1,136
No. 4	41,317	×	0.668	=	27,600
					28,974 lb
				=	29,000 lb

Straight bars (Nos. 5 & up)

No. 6	429 ft	×	1.502	=	643 lb
No. 7	1,980	×	2.044	=	4,047
No. 8	5,846	×	2.67	=	15,609
No. 9	500	×	3.4	=	1,700
No. 10	504	×	4.303	=	2,169
					24,168 lb
				=	24,200 lb

Light bending

No. 3	3,398 ft	×	0.376	=	1,278 lb
No. 4	8,607	×	0.668	=	5,750
					7,028 lb
				=	7,030 lb

Heavy bending

No. 6	2,392 ft	×	1.502	=	3,593 lb
				=	3,600 lb

Notes on the take-off (W. D. 3.5)

The take-off of the steel for the suspended slab in W. D. 3.5 illustrates the reinforcing needed in slabs and beams. Notice that the length of the rib steel is given on the longitudinal section through the rib. This is the actual total bar length (that is, the straight bar length before bending). The temperature steel in the top of the pan slab is tied with No. 2 bars every 5 ft. All the flat slabs have the same reinforcing: No. 4 on 6-in. centers in the bottom and No. 4 on 12-in. centers in the top, plus No. 4 on 16-in. centers temperature steel. Notice that W. D. 3.5 is for three slabs (including the roof).

If the lengths of the bars are specified or shown, those lengths are used for the take-off. Bar lengths not given are taken off at approximate lengths. Since it is not practical to lay out the entire reinforcing for the take-off, it must be expected that the final shop drawings will show some minor differences from the estimated quantities. For a three-floor building, the duplicated items would always be taken off either for one floor and multiplied by 3, or as one of two similar items on a floor and multiplied by 6. For example, in the corridor beams item: 1 beam is taken off, times 6 (2 per slab; 3 slabs).

The first item (spandrel beams) has a floor perimeter of 299–10; adding for 7 laps at 3–4 each, 299–10 and 23–4 gives 323–2, which is carried as 324–0. There are four No. 8 bars per beam for each of three floors, so the item is entered as $3 \times 4 \times 324$–0. The second item consists of the stirrups for the spandrel. All stirrups are measured as the beam perimeter—which in this case is twice the sum of 8 in. and 22 in., 5 ft—to allow for the hooks. A perimeter of 299–10 with stirrups on 8-in. centers will require 450 stirrups: adding one stirrup for every length of beam (that is, 6 extra) gives a total of 456 stirrups per floor. The take-off for interior beams proceeds in the same way, and is entered in the following order: the 10-in. $\times$ 14-in., 12-in. $\times$ 24-in., 10-in. $\times$ 20-in., 15-in. $\times$ 16-in., and finally 12-in. $\times$ 20-in. beams. Beams of a common length having the same reinforcing are combined for the longitudinal bars; but the stirrups are taken off separately by beam sizes. The pan slab is taken off next; it follows the detail section shown in W. D. 3.5. There are 12 pans; therefore there will be 13 joists, including the end joists at the beams. The flat-slab steel combines the two parallel slabs at the rear of the building (11–0 plus 27–6, or a 38–6 bar length).

In taking off each item, laps of 40 diameters have been allowed for every length of 30 ft or a fraction of 30 ft, and also one extra bar "for the end." For the item for temperature steel across the pan slab, the layout shown on the section through the joists is used; there are 19 bars on that section, so for both areas (only one of which is shown) there will be twice 19, or 38 bars.

Chapter 5

Structural Steel

Although structural steel is generally a subcontract item, it is often troublesome to the estimator. The structural-steel section of the specifications may include items that the structural-steel companies do not bid on; or perhaps one steel company will exclude a certain item and a competing company include it. Many structural-steel firms will exclude standard open-web joists and long-span joists. There are a great many steel fabricators who specialize in those joists and will bid only on them—either standard open-web or long-span types, or both. On very large structural-steel jobs, major steel companies will bid "supply and erect"; on smaller jobs it is more customary for them to bid "furnish and deliver," so that the general contractor must either price the erection or obtain a price from a steel erection firm.

To further complicate structural-steel bids, the specifications for that section may have included items that none of the bidders will cover—such as loose lintels, wedge inserts, the setting of anchor bolts, corrugated metal roof decking, and closure plates at metal window walls.

These miscellaneous items, plus the possibility of separate subbids for steel joists, make it necessary for the estimator to know exactly what each subbid includes and what items are required to complete the entire structural-steel work scope described in the drawings and specifications. It is also important to have the structural-steel requirements properly spelled out in the subcontracts. Thus it is often necessary for the general contractor's estimator to take off and schedule the work specified under structural steel.

In taking off the structural-steel requirements in order to check and properly evaluate steel bidders' quotations, the general contractor's estimator need not detail every item as he or she would if preparing a bid. The estimator will first break the work up into separate items such as "Structural," "Standard steel joists," and "Roof decking," and then schedule the appropriate items under each such category. The roof decking may be taken off in square feet, and the structural steel and steel joists taken off by weight. Such an approximate take-off should be reasonably complete, including all members listed and measured, but carrying only an approximate percentage figure to allow for plates, gusset plates, angle connections, and similar items. This addition for connections should vary from 2 to 8 percent of the total weight of the structural members, based on the type of job and the completeness of the structural-steel drawings. Simple one-story buildings and buildings that require steel framing for slabs may run 2 to 4 percent for connections. Multiple-story buildings that are to be entirely steel-framed (that is, all columns and floors) will usually need an allowance of from 4 to 8 percent for connections.

Working Drawing 5.1 shows the structural-steel framing for a small building. The work includes round prefabricated columns, steel beams, open-web steel joists, structural-steel lintels, and angle-iron loose lintels. The take-off will be done in detail and include all the steel items shown on the drawing.

TAKE-OFF FOR STRUCTURAL STEEL (W. D. 5.1)

H. D. Lally cols. 0–4 dia. × 10–4 (top and bottom plate) = <u>4 ea.</u>

H. D. Lally cols. 0–6 dia. × 10–2 (top and bottom plate) = <u>6 ea.</u>

Structural-steel framing

10 B 17	2	×	32–0	=	64–0	× 17 lb	=	1,088 lb
8WF 17	2	×	10–4	} 35–4	× 17	=	601
	2	×	7–4					
12WF 27	2	×	17–0	=	34–0	× 27	=	918
10WF 21	2	×	13–6	=	27–0	× 21	=	567
								3,174 lb
							=	1.6 tons

Angle-iron lintels

5 × 3½ × 5/16   3/8–0, 3/9–0, 3/9–0 = 78 ft × 8.7 lb = 679 lb
4 × 3½ × 5/16   3/5–0, 3/6–0 = 33 ft × 7.7 lb = 254
                                                        933 lb
                                              =   940 lb

Base plates & bearing plates

½ in.   8 × 8 = 8 pcs. × 9.05 lb = 73 lb
½ in.   8 × 9 = 4 pcs. × 10.2   = 41
½ in.   9 × 9 = 6 pcs. × 11.5   = 69
                                  183 lb

                              = 185 lb (18 pcs.)

Anchor bolts $\frac{3}{4}$ *in.* × *18 in.* = <u>20 pcs.</u>

Anchor bolts $\frac{5}{8}$ *in.* × *18 in.* = <u>16 pcs.</u>

Open-web steel joists

No. 147	2 × 14	=	28 at 26–6	=	742 ft	@	10.1 lb	=	7,494 lb
No. 124			13 at 17–4	=	226	@	6.0	=	1,356
									8,850 lb
								=	4.45 tons

$1\frac{1}{4}$-*in.* × $1\frac{1}{4}$-*in.* × $\frac{1}{8}$-*in. angle crossbracing*

	2 × 3 × 13 × 2 × 4–0	=	624 ft			
	2 × 12 × 2 × 4–6	=	216			
			840 ft	@	1.01	= <u>850 lb</u>

$\frac{3}{8}$-*in. dia. rod ceiling extensions* = <u>28 pcs.</u>

Bearing plates

$\frac{3}{8}$-in.   6 in. × 6 in. = 28 pcs. @ 3.84 lb = <u>108 lb (28 pcs.)</u>

Anchor bolts $\frac{5}{8}$ *in.* × *18 in.* = <u>28 pcs.</u>

1-in. × $\frac{1}{8}$-*in.* × *9-in. strap anchors*

     4(8 + 6 + 8) = <u>88 pcs.</u>

Notes on the take-off (W. D. 5.1)

Notice that all the sundries that are required for the open-web joists are taken off with those joists. This method will facilitate checking subbids that cover only either the structural or the joist items.

The round columns are a special item, and the majority of steel companies that would bid a job as small as this one would sublet the columns to a column manufacturer.

The loose lintels are kept separate from the structural-steel items because they are simple "cut-and-deliver" items involving no fabrication nor erection. The setting of loose lintels should be carried in the masonry section of the estimate.

The structural members are listed by sizes and converted to weight. For drawings on which the items are given by size only, refer to a current structural-steel data book for the weights. These data books are numerous and may be obtained free from the steel manufacturing companies.

The base plates and wall-bearing plates are converted to weight, but the number of pieces is noted also, to facilitate pricing the item. In pricing base-plate items, it is always useful to know how many pieces of plate will have to be fabricated, because the cutting will cost almost as much as the stock if the plates are small.

On the steel joist take-off, it is necessary to refer to a manufacturer's catalog to obtain the weights of the specified items. The notations "No. 147" and "No. 124" refer to the type number, standardized by the steel manufacturers. The actual average weight per foot, however, varies for steel joists according to the manufacturer's detail, because there are variations in design among manufacturers. Two particular manufacturers may each make steel joists according to the standard code, yet for joist No. 147 one maker may list 9.3 lb per ft while the other maker's No. 147 weighs an even 10 lb per ft. This variation often accounts for the different total weights given on the quotations received from different joist companies.

Miscellaneous iron

There seems to be no limit to the variety of items that find their way into the miscellaneous iron specifications. The first task of the general contractor's estimator is to separate the items that the steel bidders are not likely to include. Those items should be listed separately on the subbid sheet of the estimate, and separate prices solicited from suppliers. Items that should be segregated from the miscellaneous iron might include hollow metal doors and frames; metal toilet partitions; flagpoles; aluminum entrances; and metal sash. Having listed such items separately, it is necessary to make a checklist of the items that you expect to be covered by the miscellaneous iron bidder. When bid day arrives and the pressure is on, you may not be able to devote time to sorting out the miscellaneous iron items. At that time, you must have all the items segregated on your subsheet, in order to quickly check and evaluate the various bids.

Chapter 6

Masonry

In buildings that have relatively large amount of masonry in exterior and interior walls, the costs of masonry and masonry labor can be very extensive; consequently, it is very important that an estimator be knowledgeable in obtaining an accurate and complete quantity take-off. Quantitative masonry accounting sometimes presents problems for estimators for two reasons: first, many estimators do not understand the masonry product and secondly, masonry estimating can be, at times, a tedious task.

Many estimators find masonry quantity estimating to be difficult or tedious for numerous reasons. Sections of walls may vary from concrete masonry units, brick, glazed block, facing tile, etc., all generally occurring on a single project. The following chapter will give a good guide in reducing some of these problems and providing a quick and accurate system for the estimator. The first and most important requirement is to accept the fact that one must proceed methodically and accurately record each step of the process.

In masonry estimating, errors more commonly occur in the conversion of masonry square or cubic foot quantities to a piece count. Many estimators rely on using multiplying factors from tables as opposed to understanding the product. Many estimators use a factor that is all inclusive by attempting to include the masonry unit, header and waste. If these factors are used, one must realize that the estimate will not be accurate and that if there is a competitive bid situation, the error may be enough to either lose the bid or even worse, get the job with a reduced quantity of material and labor. In the event that subprices are received, there still needs to be an accurate quantity take-off to ensure that problems are not encountered with the subcontractor at a later date.

Throughout the country, there are many different types of masonry materials that are manufactured in a great variety of sizes. Face brick, as an example, will average $1\frac{1}{8}$ in. to $2\frac{1}{4}$ in. in height, shale brick will average $2\frac{3}{8}$ in. in height, glazed brick in the standard nominal size of $2\frac{3}{8}$ in. $\times$ 8 in. is actually $7\frac{3}{4}$ in. long with a "permissible variation" of $\frac{3}{32}$ in. height and $\frac{3}{16}$ in. length. "Permissible variation" according to the Facing Tile Institute is the maximum difference or variation allowed between the smallest and largest units in any one order. Hardburned dense face brick that is classified as standard size (8 in.

× 2¼ in. × 3¾ in. nominal) will average $7\frac{5}{8}$ in. length, while other standard size face brick will average 8 in. to $8\frac{1}{16}$ in. length. In computing brick quantities, the height of the brick is not usually a questionable factor. The coursing (brick plus joint) as shown on the drawings or in the specification will control the height. The length, with its many variations, and the fact that architects rarely lay out buildings according to modular masonry dimensions are the problem creators. Other problems are that standard brick from different plants may vary $\frac{3}{8}$ in. or more, common brick, unlike face brick, is not culled out, causing problems when face brick is to be backed up by common brick.

For the past 40 years, individuals dedicated to the coordination and standardization of building materials have argued to develop and promote the modular system of design and manufacture. The standard module is a 4-in. unit in all three dimensions, length, height, and thickness. As it affects masonry, the modular standard requires that brick and joint conform to the 4 in. unit, bricks are to be a length that will lay 8 in. or 12 in. (including the joint) and of a height that will provide a vertical rise of 12 in. for every three courses. The 5-in. facing tile in modular layout would course $5\frac{1}{3}$ in. or three courses per 16 inches. Proponents of the modular standards contend that the system reduces both cost and time. That would be true only if modular standards were universally accepted.

The modular system has made great progress. Today there are several brick manufacturers who are making modular-sized material exclusively. There are also hundreds of brick manufacturers who are converting their plants to the production of modular-sized material.

In the gradual conversion to modular standards, the architect is a key figure. No amount of standardization of materials will accomplish anything until the architects and engineers bring their designs into conformity. Architects are nonconformists and are not going to be bound by standards and limitations that they feel would restrict their design capabilities. The immediate alternative remains for the estimator to know the product, be methodical, and keep good records.

PRELIMINARY ANALYSIS

In computing any masonry quantities, several factors must be determined before the square-foot area can be converted into pieces of brick, facing tile, concrete block, etc. Vertical and horizontal coursing is used to compute the net number of pieces per square foot, then add the headers and waste. The height of the masonry unit is not the major concern, if five courses are required in $13\frac{1}{8}$ in., then a coursing of $2\frac{5}{8}$ in. must be used for estimating the masonry unit quantities. The height of the brick, whether $2\frac{1}{4}$ in. or $2\frac{3}{8}$ in., does not matter, since the difference will be made by the thickness of the mortar joint.

Because masonry products do have considerable variations due to the nature of clay materials used, the time that the units are baked or cured, and the temperature of the kiln during curing, one must first determine the actual length of the masonry units to be used. This can be determined by examining a sample strap of the material or by contacting the supplier. A difference of $\frac{1}{4}$ in.

from the average length between different types of brick could mean a difference of 3 percent in quantity of material needed to be purchased and installed.

Another important factor in computing masonry quantities is the required bonding. Bonding that requires the use of headers, as in a cavity wall, means that the masonry unit must be turned and extended inward into the masonry backup wall. The interior wall could be common brick or concrete block. The use of concrete block with headers will require either block and common brick or concrete brick filler pieces between the header units. The header units calculated in the exterior face of the masonry wall will need to be deducted from the interior face since they protrude into the interior face.

Masonry waste varies depending on the type of bond and material. For example, buildings having long stretches of straight walls with headers and few openings have considerably less waste than buildings with short stretches, offsets, and numerous openings. The waste factor will vary in the range of 3 to 5 percent. Walls that have snapped headers (headers that are broken and do not extend inward into the interior wall surface) will afford the use of many culled units since only half of the masonry unit is used. Caution must be used when planning to use the culled or waste materials. First, the snapped masonry unit will often chip and be unusable and second, the masonry contractor must make a conscientious effort to use as much of the culled material as possible. The use of snapped headers will usually yield waste factors near 5 percent. The type of material used will also contribute to waste. Hard deeply burned masonry units that produce the darker richer colors tend to be more brittle and chip or spall more than some other masonry products. Examine the material and get all the information available from manufacturers and masonry contractors that may be available.

ESTIMATING BRICK QUANTITIES

We will consider the estimating face brick quantities first. Face brick is generally more expensive and quite often involves some type of masonry bond. The quantity take-off, which is indicated in square feet, will be shown in examples later in this chapter. In computing the face brick quantities, we will consider the total net area 13,006 SF for a 12-in. wall that is made up of 4 in. of face brick and 8 in. of common brick backup. The coursing will be $2\frac{5}{8}$ in. (brick plus mortar joint) and the length will be $8\frac{1}{8}$ in. ($7\frac{3}{4}$ in. plus $\frac{3}{8}$ in. of mortar) for both the exterior wall and interior backup wall. The exterior face brick will be laid with a Flemish header bond every sixth course. The calculation for the number of bricks per square foot would be:

$$= \frac{12 \times 12}{8\frac{1}{8} \times 2\frac{5}{8}}$$
$$= 12 \times 12 \times \frac{8}{65} \times \frac{8}{21}$$
$$= \frac{9,216}{1,365}$$
$$= 6.75 \text{ per SF}$$

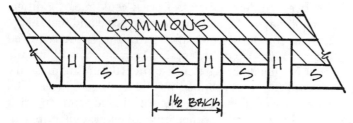

Figure 6.1 Face brick. Flemish-bond course.

The Flemish bond course is shown in Fig. 6.1. In a panel of brick that is $1\frac{1}{2}$ bricks wide and 6 courses high, there are 9 bricks on the face plus one-half of the header, which does not extend back and into the backup space. There are nine half-face header brick shown; therefore, an allowance for headers in the backup space should be also $\frac{1}{2}$ unit per 9 units or $\frac{1}{18}$ of the net face quantity. This would be shown as follows:

Exterior face brick

$$
\begin{array}{rl}
\text{Total area} & = 13{,}006 \text{ SF} \\
8\frac{1}{8} \times 2\frac{5}{8} = 6.75 \text{ per SF} \times & 6.75 \\
\hline
& 87{,}790 \text{ pcs.} \\
+ \text{ Headers } \tfrac{1}{18} & 4{,}877 \\
\hline
& 92{,}667 \\
+ \text{ Waste } 3\% & 2{,}780 \\
\hline
& 95{,}447 \text{ pcs.} \\
= & \underline{95.5 \text{ M}}
\end{array}
$$

Common-brick back-up

$$
\begin{array}{rl}
\text{Total area} & = 13{,}006 \text{ SF} \\
8 \text{ in.} \quad 2 \times 6.75 = 13.5 \text{ per SF} \times & 13.5 \\
\hline
& 175{,}581 \text{ pcs.} \\
\text{Less Headers} & 4{,}877 \\
\hline
& 170{,}704 \\
+ \text{ Waste } 3\% & 5{,}121 \\
\hline
& 175{,}825 \text{ pcs.} \\
= & \underline{176 \text{ M}}
\end{array}
$$

Similarly, the common brick wall surface will be computed by taking off the entire 12-in. wall quantity and deducting the face brick quantity.

You will see that the final number of pieces has been rounded off and that the adjustments are relatively small because of the total quantity involved.

Common-brick back-up

Total wall area (12 in.)					=	13,006 SF	
12 in.	3	×	6.75	= 20.25	×	20.25	
						263.371 pcs.	
				+ Waste 3%		7,901	
						271,272	
				Less Face brick		95,447	
						175,825 pcs.	
					=	176 M	

The coursing used on the exterior face was $2\frac{5}{8}$ in. vertically and $8\frac{1}{8}$ in. horizontally with face-brick headers. The interior backup must maintain the same vertical coursing; however, an estimator who is familiar with its masonry subcontractor, could potentially increase the horizontal length by increasing the end masonry joint to $\frac{1}{2}$ in., producing a brick plus joint length of $8\frac{1}{4}$ in. An estimator who is alert and knowledgeable could potentially create a more competitive bid. Caution should be taken and the estimator's plan should be noted and passed onto the project manager in a turnover meeting from estimating to construction. Further, one must determine if the design architect will accept the wider joint on the backup surface.

Common-brick back-up

Total area (as before)					=	13,006 SF
8-in. wall	2	×	6.65	= 13.3 per SF	×	13.3
						172,980 pcs.
				Less Headers		4,877
						168,103
				+ Waste 3%		5,047
						173,150 pcs.
					=	173.15 M

If there are no objections to using the wider joint, the backup brick quantities in the previous example could be taken off with the common brick coursing of $8\frac{1}{4}$ in. × $2\frac{5}{8}$ in. or 6.65 pieces per square foot. Using a $2\frac{3}{4}$ in. × $8\frac{1}{4}$ in. coursing, the conversion factor would be:

$$12 \times 12 \times \frac{4}{11} \times \frac{4}{33} = \frac{2,304}{363}$$

$$= 6.35 \text{ per SF}$$

Using a coursing of $2\frac{1}{2}$ in. × 8 in., the conversion factor would be:

$$12 \times 12 \times \frac{2}{5} \times \frac{1}{8} = \frac{36}{5}$$

$$= 7.2 \text{ per SF}$$

It should be noted that although the above examples are for standard brick that are nominally 8 in. × 2¼in. × 3¾ in., the number of brick varies from 6.35 to 7.2 per square foot due to differing vertical and horizontal coursing.

A *full Flemish bond* with header and stretcher alternating in every course requires two bricks (1 header and 1 stretcher) to cover the area of one and one-half bricks on the face; therefore, the addition for headers is a half brick per one and one-half brick, or one-third. Therefore, the quantity for headers will be one-third of the net quantity.

A *full English bond* with alternating stretcher and header requires two stretcher and one header or three bricks per each bond; therefore, add one-half brick for each header.

A *full header bond* every fifth course can be determined by examining a panel one brick wide with four courses of stretchers and one course of two headers requiring six bricks to show five or an addition of one in five or ⅕. For this type of header bond add one-fifth of the total net units for headers. (See Fig. 6.2.)

A Flemish header bond every fourth course has three stretcher courses and one header course alternating. In a four-course panel, one and one-half bricks wide, 6½ bricks are required to show six or ¹⁄₁₂. Add for headers of this type one-twelfth of the total net units for headers.

The requirement for any bond may be computed in a similar manner. One must determine how many bricks are to be laid and how many shown in the face area. The addition for headers is the quantity of face brick within the backup area expressed as a fraction of those that show on the face. The total backup area is usually different from that of the face; therefore, the actual number of pieces of face brick within the backup area must be computed in order to make the proper deduction of face brick from the gross backup brick quantity to obtain the actual backup brick quantity. Using an all-inclusive

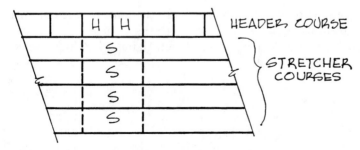

Figure 6.2 Face brick. Full header bond every fifth course.

multiplying factor will not give the estimator an accurate quantity take-off. If one is to be competitive as well as providing an adequate count for construction, the estimator must make an accurate take-off.

Concrete block backup

When using concrete block as a backup, special considerations must be given to the different type headers, for example:

1. A 12-in. wall with 4 in. of face brick using a Flemish header in the sixth course and 8 in. of concrete block backup (Fig. 6.3).

   The concrete block shoe block or header block is recessed to receive the headers. The recess is continuous for the entire length of the concrete block, which can be either 16 or 18 in. long. Because of this shape and continuous recess for the header extension, the space on either side of the header must also be filled in with either common or concrete brick. As can be seen, there will be one common brick filler required per header, or twice as many common bricks as were added to the face-brick headers.

2. The next example is a 12-in. wall with 4 in. of face brick and a full header course every fourth course and 8 in. of concrete block backup. Common brick are shown as the backup in the header course (Fig. 6.4).

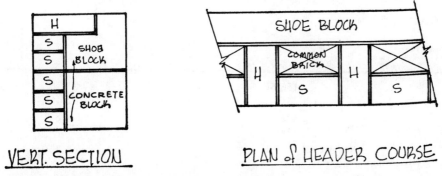

Figure 6.3 Face brick. Flemish bond every sixth course.

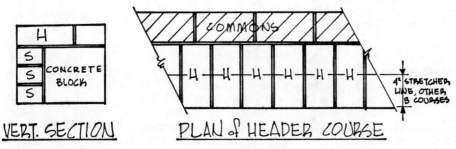

Figure 6.4 Face brick. Full header bond every fourth course.

The header brick uses the entire center 4 in. of the wall in the header course, with common brick backup for the remaining inner 4 in. The quantity of common brick required is the same as the quantity of face brick that is added for the headers. The concrete block will be 75 percent of the face-brick area. If deductions are made from the concrete block backup for columns, spandrel beams, etc., then one can subtract the concrete block backup deductions from the face-brick area to arrive at the area for the concrete block computations.

SINGLE-STORY BUILDING

Working Drawing 6.1 shows a simple masonry building with a 12-in. masonry exterior wall, brick shelf at the foundation wall, and concrete spandrel beams. The masonry take-off is straight forward, but contains almost everything that would be found in any major building. A multistory building several hundred feet long would be only slightly different, having perhaps concrete columns, different sizes of window openings, and varying sizes of spandrel beams. Once the basic method of taking off the quantities has been mastered, the size of the project should be of no consequence.

The area of brick shelf and the concrete spandrel beams are included in the concrete take-off. These areas are also shown here: brick shelf equals 156 SF and spandrel face area equals 165 SF.

The take-off (W. D. 6.1)

MASONRY

Ext. waterstruck face brick

```
Shelf                           =    156 SF       Outs
132-0  ×  9-7                   =  1,265          Drs.   2/3-6 × 7-0  =   49
                                   ─────          Wind.  3/6-0 × 4-2  =   75
                                   1,421                                ─────
                                −    124                                124 SF
                                   ─────                                ═════
                       Net Area  1,297 SF
8⅛ × 2½ = 7.09 per SF    ×        7.09
                                 ──────
                                 9,195 pcs.
Headers on
976 SF      1/18 × 6,920   +      385
                                 ──────
                                 9,580
             + Waste 4%           383
                                 ──────
                                 9,963  =  10 M
                                 ═════
```

Common-brick back-up

```
                  2  ×   385 pcs.  =   770 pcs.
                      + Waste 4%        30
                                      ──────
                                       800 pcs.  =  0.8 M
                                      ══════
```

8-in. concrete-block back-up

Total face area			=	1,297 SF
Less Shelf	156 SF			
Spandrel	165			321
				976 SF

16 in. × 7½ in. = 1⅕ per SF + ⅕   195
                                                       1,171 pcs.
                                   + Waste 4%   49
                                                    1,220 pcs. (8 × 16)
                                                             (Note—7½-in. coursing)

Wash down face brick = 1,297 SF

Ext. scaffold = 1,500 SF

Int. scaffold
              125–4 × 8–4 = 1,044 SF

Clean and point concrete block = 920 SF

Notes on the take-off (W. D. 6.1)

The face-brick take-off starts with the brick shelf area, which is obtained from the concrete sheets. The 132–0 perimeter is obtained by adding dimensions of 33–7 plus 28–8 plus 3–9 doubled. The 33–7 represents the outside of the face brick and the 28–8 dimension represents the inside dimension of 29–4 less twice 0–4. The openings are then deducted to obtain the net face-brick area. The brick coursing works out to be 7.09 pieces per SF; therefore, 9,195 units represents the net quantity of face brick required for the face wall. There are no headers to add for the concrete backup; therefore, the net face-brick area for headers is 1,297 SF less the sum of 156 SF plus 165 SF for a net total of 976 SF, which yields 6,920 units. Flemish headers in the sixth course add one-half brick per each nine face bricks or one-eighteenth of the net face-brick area for the headers. Some headers will snap and chip requiring a waste factor. Three percent waste would probably be tight; therefore, 4 percent would be a more reasonable waste factor to use.

    A Flemish header course would need one full common brick filler between each header of one full filler piece for each half header, or twice as many commons as the number of face bricks for headers.

    The concrete-block backup area is the same as the face-brick area. The actual backup area is less than the face-brick area by the difference in the four corners; however, concrete-block backup in the corners is not deducted because the waste will be considerable due to the block length. The concrete block is nominally 16 in. long, with some manufacturers making an 18-in. (nonstandard) block. The concrete block in this example must course 7½ inches to match the three brick courses, which requires the use of a 7⅛-in. block. Note that this may require a special order for the block since most concrete masonry units are 7⅝ in. high, which courses with three brick courses at 8 in.

In our example one block plus the joint dimension is 16 in. × 7½ in., which equals 120 sq. in. or ⅝ SF. This results in 1⅕ block per SF; therefore, the square foot area should be increased by ⅕ to convert to a piece count. Where convenient, one will typically adjust the waste factor to round out the final totals. The shoe blocks are usually shown separately only if they differ in price from an ordinary block or if the take-off was for purchasing rather than estimating costs.

Concrete block 16 in. long including joint and coursing will cover 128 sq. in. or 8/9 SF. This yields 1⅛ block per SF, requiring the total area to be increased by ⅛ to convert to a piece count.

Several miscellaneous items must be considered in the masonry estimate:

The area to be cleaned will be the net area of the masonry surfaces.

Scaffolding is estimated by the gross area for the exterior face-brick and interior backup masonry area.

The cleaning and pointing for concrete block is equal to the interior net area of masonry surfaces.

The conversion from square feet to a piece count in concrete block may be worth extra attention, since it seems to give trouble and cause arguments. If the block plus the joint is 16 in. long by 7½ in. high, then it covers 129 sq. in. One square foot equals one-fifth more area than one block face surface; consequently, one must add one-fifth more to convert from square feet to a piece count.

In the above masonry take-off for W. D. 6.1, one can begin to see several items emerging that will improve the accuracy and the speed in making the estimating take-off. First, there is the value of retaining estimating work sheets. As an example, from the concrete work sheet, you will note that deductions made from the masonry were based upon the previous concrete take-off. By keeping the work sheets, the estimator can save time by not having to repeat take-off calculations for similar areas. Second, by being methodical and accurate in the initial calculations, the estimator obtains an accurate quantitative take-off of the materials. Next, one can see that much of the process can be simplified by making fewer take-offs, as in the conversion of the face-brick area to the masonry backup area. As previously stated, it does not matter what material is being used for the masonry backup because conversions can be made. Therefore, we are only concerned with making the proper deductions for openings, columns, and beams, etc.

The following are further examples of masonry take-off procedures. The methodology remains the same as previously discussed.

Common-brick backup

Consider the building in W. D. 6.1. If the backup is to be 8 in. of common brick, the face-brick quantity would be unchanged and the common-brick backup would be as follows:

Common-brick back-up

8 in.	128–0	×	8–4	=	1,067 SF
	Less Outs			=	124
					943 SF
			×		14.18
					13,372 pcs.
	Less Headers				385
					12,987
	+ Waste 4%				513
					13,500 pcs.
				=	13.5 M

For the common brick backup, the four corners are deducted. Each corner is 1 ft less than the face brick (Fig. 6.5) resulting in the common-brick perimeter of 132–0 less the four 1-ft corners. The common-brick height is 9–7 less the spandrel beam depth. The openings must be deducted from the area, similar to the face-brick deductions. The common-brick backup area could also have been obtained by deducting the four corners from the backup face area:

$$(976 \text{ SF}) - (1\text{–}0 \times 8\text{–}4) \times (4) = 943 \text{ SF}$$

The common-brick backup is 8 in. thick. Using the conversion factor found for the face brick will yield (2) (7.09) units per SF, or 14.18 pieces per SF. The headers are added to the face brick count for the net quantity before waste. Therefore, this must be deducted from the common brick totals before calculating the waste quantity.

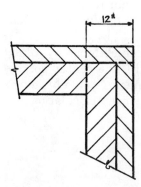

Figure 6.5 Backup brick at corners.

FOUR-STORY BUILDING

Working Drawing 6.2 indicates a four-story building. The first and second floor exterior walls are 16-in. masonry, the third and fourth floors have 12-in. masonry walls. Additional information from the drawings is as follows:

Schedule of openings

Heating louver (1st & 2nd)	1 at 32 SF			*Brick shelf* =	1,807 SF	
Doors (1st & 2nd)	3 at 6–8	×	7–3½			
Doors (1st & 2nd)	7 at 3–6	×	7–3½	Concrete columns	1st	560 SF
Sash A (1st)	4 at 18–2	×	5–4	Concrete columns	2nd	536
Sash B (1st & 2nd)	28 at 4–3	×	4–7	Concrete columns	3rd	500
Sash C (1st)	20 at 4–3	×	5–4	Concrete columns	4th	410
Sash D (2nd)	20 at 4–3	×	3–8			
Sash E (1st)	2 at 10–6	×	5–4	Concrete spandrels	2nd	1,610 SF
Sash B (3rd)	28 at 4–3	×	4–7	Concrete spandrels	3rd	1,610
Sash E (3rd)	16 at 10–6	×	5–4	Concrete spandrels	4th	1,042
Sash D (4th)	36 at 4–3	×	3–8	Concrete spandrels	roof	744

Shale face brick—2¾-in. coursing; full header course every 3rd course.
Concrete block back-up—shoe blocks throughout (for headers in 3rd course).

The collection sheet (Table 6.1) should be similar to the one shown for a building structure of this size.

The perimeter is taken off for each of the three floor plans at the first, third, and fourth floors. The 16-in.-thick wall extends up two floors for a height of two times 11–11 or 23–10. The exterior dimension is a rectangle 281–4 × 108–0 times two plus 26–6 times four for the setbacks (Fig. 6.6). The third-floor perimeter can be quickly checked and is equal to the first-floor perimeter less 40–10 times two (883–4 − 81–8 = 801–8) the fourth-floor perimeter is a simple rectangle.

The 12-in. thick masonry wall take-off could have been obtained by taking off the third-floor plan at the end walls (10–6½ high), the center portion (21–1) and the fourth-floor end walls (10–6½). This sequence would have provided the same total as the earlier method, with slightly more calculations.

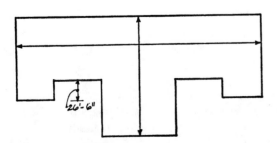

Figure 6.6 Perimeter for exterior masonry.

TABLE 6.1   Collection Sheet for Exterior Masonry

				Outs		Conc. (SF)	
Ext. masonry — face brick			Drs.		Sash	Cols.	Span.
		23–10	10–0 × 9–2	4/18–2 × 5–4	560	1,610
4-in. face		281–4	3/6–8 × 7–3½	28/ 4–3 × 4–7	536	1,610
+		67–2	7/3–6 × 7–3½	20/ 4–3 × 5–4	1,096	3,220
12-in. blocks		40–2		20/ 4–3 × 3–8		
	2/26–6	53–0		2/10–6 × 5–4		
		441–8				
		2				
		883–4				

				Outs		Conc. (SF)	
				Sash		Cols.	Span.
4-in. face		10–6½	10–6½	28/ 4–3 × 4–7	500	1,042
+		281–4	204–4	16/10–6 × 5–4	410	744
8-in. blocks		93–0	66–6	36/ 4–3 × 3–8	910	1,786
		26–6	270–10			
		400–10	2			
		2	541–8			
		801–8		Shelf	1,807	
				Cols.	2,006	
		(1,343–4)		Span.	5,006	
					8,819 SF	

The take-off (W. D. 6.2)

EXTERIOR MASONRY

Ext. shale face brick   OUTS

```
            Shelf        =  1,807 SF   Heating louver        =    32 SF
 (4 + 12)   883–4 × 23–10  = 21,053   Drs.   3/ 6–8 × 7–3½ ⎫
 (4 +  8) 1,343–4 × 10–6½  = 14,161   Drs.   7/ 3–6 × 7–3½ ⎬       325
                             37,021   Sash   4/18–2 × 5–4  ⎫    16-in.
                          −   4,169   Sash  20/ 4–3 × 5–4  ⎬  953  wall
                Net area    32,852 SF Sash   2/10–6 × 5–4  ⎭    2,167
 8¼ × 2¾ = 6.35 per SF   ×    6.35    Sash  28/ 4–3 × 4–7  =   545
                           208,610 pcs. Sash 20/ 4–3 × 3–8 =   312
 + Headers ⅓ × 152,610  =   50,870
   32,852                  259,480   •Sash  28/ 4–3 × 4–7  ⎫    12-in.
 −  8,819      + Waste 4%   10,380    Sash  36/ 4–3 × 3–8  ⎬ 1,106  wall
   24,033                  269,860 pcs. Sash 16/10–6 × 5–4 =   896
 ×  6.35                 =     270 M                          4,169 SF
  152,610
```

12-in. concrete-block back-up

Total face area			=	21,053 SF		
Less Opgs.	2,167 SF					
Cols.	1,096			6,483		
Span.	3,220					

$$16 \times 8\tfrac{1}{4} = 1\tfrac{1}{11} \text{ per SF} + \tfrac{1}{11}$$

```
                                          14,570 SF
                                           1,325
                                          15,895 pcs.
                              + 4%           635
                                          16,530 pcs.  (8 × 16)
```

8-in. concrete-block back-up

Total face area		=	14,161 SF	
Less Opgs.	2,002			
Cols.	910		4,698	
Span.	1,786			

```
                          + 1/11    9,463 SF
                                      860
                                   10,323 pcs.
                          + 4%        417
                                   10,740 pcs.  (8 × 16)
```

Wash down face brick = 32,852 SF

Exterior scaffold = 37,100 SF

Interior scaffold

$$2 \times 879\text{-}4 \times 10\text{-}1 = 17{,}733 \text{ SF}$$

```
            797-8 × 9-2 }
            537-8 × 9-2 }  12,241
                          ─────────
                          29,974 SF
```

Clean & point concrete block = 24,033 SF

Notes on the take-off (W. D. 6.2)

With the collection sheet properly prepared, the transfer of data should be accomplished quickly. The 16-in. walls are separate from the 12-in. wall and the two floors at 10–6½ are added together to simplify the brick calculations. Note on the collection sheet that brick against concrete areas are added for a total area of 8,819 SF of face-brick which will not require a backup material. The 8,819 SF is deducted from the total face-brick area, leaving 24,033 SF to be backed up with concrete block. The headers that are required against the concrete wall will be snapped. A full course of headers every three courses requires four bricks to show three on the face. This results in an addition of one in three or one-third for headers.

The deductions are subtracted to reduce the calculations by keeping common dimension items together. The doors are an example resulting in a 325 SF deduction.

The concrete-block backup is obtained by subtracting the concrete columns and concrete spandrel beams that are located in the masonry, from the gross face-brick area (21,053 SF) located on the second line of the face-brick quantities. The result is the net area of 12-in. concrete-block backup required. The brick courses $2\frac{3}{4}$ in., which is $8\frac{1}{4}$ in. for three brick or one concrete block. The block plus a joint will be 16 in. by $8\frac{1}{4}$ inches or 132 sq. in. One square foot of area will contain $1\frac{1}{11}$ blocks; therefore, one-eleventh must be added to the square foot area to convert to a piece count. If 12-in. concrete-block headers are not available from local manufacturers, the take-off will require the use of 8-in. header block and 4-in. standard block.

The exterior scaffolding will be the gross brick area (37,021 SF) plus a small allowance for the distance from the ground to the bottom of the brick work. If the grade around the building is low, the item for rubbing the exterior foundation walls will provide the additional area to be allowed in computing the exterior scaffolding.

The interior scaffolding area is equal to the inside perimeter times the brick height from the floor to the underside of the spandrel for each floor.

The concrete block will be exposed throughout; consequently, an item for cleaning and pointing is required. If the concrete block were to be exposed only in certain areas, this item could be priced to include an extra cost for laying as well as cleaning and pointing.

For W. D. 6.2, with solid common-brick backup, the face-brick quantity would be similar to the above. The backup take-off would read as follows:

Common-brick back-up

12-in.	Face area		=	21,053 SF			
Less	Opgs.	2,167	}	6,483			
	Cols.	1,096					
	Spand.	3,220					
				14,570 SF	× 12 in.	=	14,570 CF
8-in.	Face area		=	14,161 SF			
Less	Opgs.	2,002	}	4,698			
	Cols.	910					
	Spand.	1,786					
				9,463 SF	× 8 in.	=	6,309
							20,879 CF
						×	19.05
							397,745 pcs.
				Less Headers			50,870
							346,875
				+ Waste 4%			13,875
							360,750 pcs.
						=	360.75 M

EXTERIOR WALLS WITH PREFORMED WATERPROOFING

There are several waterproofing companies producing preformed fabric waterproofing for use on exterior walls. Figure 6.7 shows a typical exterior wall with preformed waterproofing. The backup area will consist of 50 percent 8-in. blocks and 50 percent 4-in. blocks with common-brick fillers. The common-brick item will be 50 percent of the backed-up area, less the add for face-brick headers extending into the backup area.

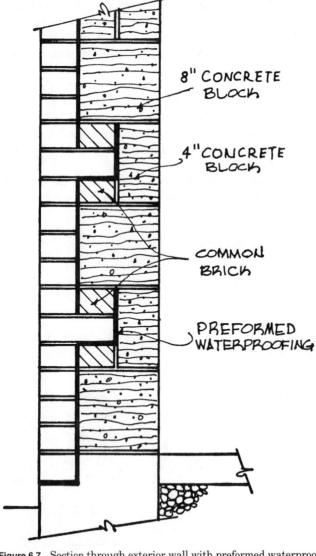

Figure 6.7 Section through exterior wall with preformed waterproofing.

INTERIOR MASONRY PARTITIONS

Masonry partitions may be any one of many materials: brick, concrete block, gypsum block, clay tile block, or structural facing tile. Unless the drawings show otherwise, all partitions should be taken off to the underside of the slab or beam. For a large building that has several types of partitions of varying heights, it is advisable to mark the partition information on the plans. First mark the heights of the partitions that vary from the predominant height (for the floor being taken off) alongside each one on the plan, and then mark that predominant height clearly in colored pencil, for example, "Ptns. 9–6 except noted." Next, in the applicable rooms, mark the materials that vary from the predominant material. If all interior partitions were to be of concrete block except a few of structural facing tile, then it would only be necessary to mark the areas requiring facing tile.

When all the required information has been transferred onto the floor plans, then the partitions can be taken off without having to turn from one sheet of the drawings to another. Constantly referring to the various drawings and room schedules while trying to take off a particular floor wastes a great amount of time. If all floor plans are properly noted before the take-off is begun, it should be possible to see where materials and room layouts repeat, saving time by making the repeating items work for you.

Interior partitions generally are straightforward items and except for structural glazed material, no difficulties should be encountered. The most common masonry materials used for partitions are described below.

Concrete block typically has nominal face sizes or 8 in. × 16 in. and 8 in. × 18 in. and thicknesses of 2, 3, 4, 6, 8, 10, and 12 in. The actual length and width are approximately $\frac{3}{8}$ in. less than the nominal dimensions. The block height will be one of the following (rising one block to three brick courses): $7\frac{1}{8}$-in. block for $7\frac{1}{2}$-in. coursing; $7\frac{1}{2}$-in. block for $7\frac{7}{8}$-in. coursing; and $7\frac{5}{8}$-in. block for modular 8-in. coursing. Not all manufacturers make blocks of all three heights, so the requirements should be checked with the local plants.

Gypsum block in. × 30 in. (2, 3, 4, and 6 in. thick).

Load-bearing clay tile (T. C. tile)—12. in. × 12 in. (4, 6, 8, and 12 in. thick).

Structural facing tile is usually referred to by the following designation: 4S series, $2\frac{3}{8}$ in. × 8 in.; 4D series, 5 in. × 8 in.; 6T series, 5 in. × 12 in.; 8W series, 8 in. × 16 in. All sizes are nominal face dimensions (height and length). Facing tile stretchers are available in 2-, 4-, 6-, and 8-in. thicknesses, but not all series are made in all thicknesses. There are also numerous special shapes. In fact, facing tile cannot be taken off properly without using the manufacturer's catalog. Facing tile is made in several finishes: unglazed, salt-glazed, clear-glazed, and ceramic-glazed.

Glazed concrete block sizes and thickness are the same as for ordinary concrete blocks, except that the 3-in. and 10-in.-thick blocks are not standard glazed items. Specials include bullnosed jamb and head pieces, coved-base blocks, square quoins, and miter blocks. As for facing tile, a manufacturer's catalog should be used to identify the various specials.

Working drawings 6.3 and 6.4

Working Drawings 6.3 and 6.4 show the interior masonry partitions and the room finish schedules. A considerable amount of information is given, much of it descriptive information that would ordinarily be given on other drawings or in the specifications. Some of the notations on the plan are our markings, such as the "G.F.T." notations, the window sizes, and the partition heights. Note that where a 6-in. partition requires a glazed facing tile dado with plaster above, the facing tile has a bullnosed cap and the partition above is 4 in. thick. These requirements are necessary because the 6-in. facing tile partition will measure $5\frac{3}{4}$ in. and the 4-in. block $3\frac{3}{4}$ in.; therefore, the partition, plastered on both sides, will be 5 in. wide ($3\frac{3}{4}$ in. plus $1\frac{1}{4}$ in.), and $\frac{3}{8}$ in. of bullnose will be exposed. The alternative construction would be a 4-in. partition with a full inch of plaster on each side, so that the plaster would be flush with the facing tile. (See Table 6.2.)

The collection sheet is accomplished by taking off the facing tile areas first and progressively adding the concrete-block items to complete those walls and then taking off the walls that are entirely concrete blocks. The facing tile is measured as if it were stretchers and the specials are taken off as extra-cost items. The facing-tile specials are taken off by group regardless of type, because all pieces in a particular group will be the same price (Group 1 to Group 6). You will need a manufacturer's booklet to follow the facing tile take-off.

The "outs" for doors were taken off as "Singles" or "Pairs." In estimating masonry partitions, door openings should be deducted at 18 SF for a single door and 35 SF for a pair, unless the openings are abnormally large. In buildings (such as hospitals) that have a considerable number of 4-ft single doors, the 4-ft doors should be taken off separately and deducted at 25 SF per single opening.

Window openings in the 2-in. furring tile for the exterior walls are deducted to the nearest tile multiples (even feet for the length and the tile coursing for the height) less than the openings; that is, a 21-ft 9-in. opening would be deducted at 21 ft, not 22 ft.

In Room No. 2 Window Wall (Fig. 6.8), the 2-in. glazed facing tile is taken off at 36 ft long, with an "out" of 21 ft × 10 courses for the window. The 8-in. west wall (26–6) around the corner to the corridor (3–6) is 30–0 in length less an "out" of 1 door for 22 courses of 2-in. glazed facing tile with 6-in. concrete block backing 9–8 in height; above this, the wall requires 8-in. concrete block 1–6 high, taken off to the full 30–0 in length. The south wall of Room 2 to the corridor is 32–8, with 4-in. glazed facing tile on both sides: 22 courses high on the room side and 12 courses on the corridor side; 1 pair of doors (marked "C") is deducted from the room side and for the corridor side the same door is deducted at 25 SF. Above the 12-course facing tile on the corridor side is 3-in. concrete block, 4–4 high at 32–8 in length. This completes the take-off for the

TABLE 6.2 Collection Sheet for Interior Masonry

G.F.T. 4-in. (clear)				G.F.T. 4-in. G2F				Ceram. F.T. 4-in.	
22c	12c	20c	Outs	12c	16c			19c	
32–8	32–8	41–0	Drs. S 1√	9 ft√	8 ft√			10–0	
√	21–0	√						16–0	
	12–0		Drs. pr. 1√	Clear G.F.T. specials				26–0√	
	33–0								
	33–0		Drs. 25 SF 1√	1	2	3	4		
	26–0							Ceram. F.T. 2-in.	
	41–0		Drs. 14 SF 2	22	21	8	4		
	31–0		1	32	21	11	4	19c	Out
	229–8		2	24	14	6	2	26 ft√	Drs. 2√
	√		√5	48	17	15	4		
			=	24	17		2		
				22				Ceram. F.T. 4-in. G2F	
Opg. 1/6–0 × 5–3√				172	⎰66	42	6	19c	
				Base	⎱33	45	10	10–0√	
G.F.T. 2-in. (clear)					⎰41	38	8		
					⎱83	144	14		
22c	12c	20c	Outs						
36–0	41–6	22 ft	Sash	Cap	60		6	8-in. concrete block	
30–0	41–4	19	21–0 × 10c√		227		14	1–6	11–2
5–6	15–6	41 ft	17–0 × 10c√		600	309	74	30–0√	16–0√
21–0	98–4	=			√	√	√		
92–6	√		Drs. 1 + 1√						
√			Dr. pr. 25 SF√	3-in. concrete block					
	Opg.	6 × 7 ft = 1√		4–4	3–0	3–6			
	Dr.	14 SF = 1		32–8	26–0	41–0			
			3	21–0	√	√			
			4√	53–8√					
			=						

4-in. concrete block					6-in. concrete block			
9–8	1–6	6–0	11–2	Out	9–8	1–6	11–2	Out
5–6	21–0	12–0	14–6	Dr. S 1	30–0	32–8	41–6	Drs. 1
√	√	41–4	12–6	3	√	5–6	38–0	1
		31–0	21–0	2	38–2	79–6		2√
		√84–4	48–0√	1		√	√	=
				7√				
				=	6–0	2–5		
2–4	3–0	8–9	5–3		33–0	41–0		
10–0	26–0	19–0	41–4			19–0		
16–0	√	√	√		√	60–0√		
10–0								
36–0√								

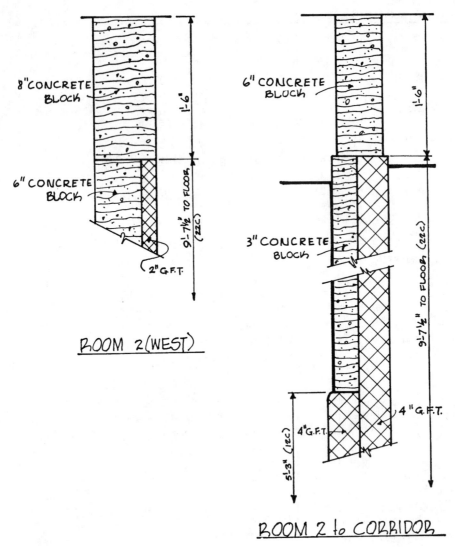

Figure 6.8 Details for interior masonry.

wall up to 9–7½ (22 courses) in height. We take off 6-in. concrete block, 1–6 high at 32–8 to pin up to the slab.

The take-off continues item by item, completing one area of facing tile before proceeding to another area. After all walls requiring facing tile have been completed, the facing-tile specials are taken off.

The Group 1 items are: window jambs for Room 2 (2 times 11 pieces) 22 pieces, for opening E (2 times 16 pieces) 32 pieces plus (2 times 12 pieces) 24 pieces, for Room 13 (2 times 11 pieces) 22 pieces. Quoins for corridor (4 times 12 pieces) 48 pieces and for door "X-1" (2 times 12 pieces) 24 pieces. Total: 172 pieces.

Group 2 items are the window sills and heads. The head at opening E is 2-in. coved base and dado caps.

Group 3 items are 4-in. double-bullnosed caps for the low partitions (each 1 piece less than the full partition length), the double-bullnosed jamb pieces for the low partitions (each 1 piece less than the full height) and 4-in. coved base.

Group 4 items are four mitered pieces at the window in Room 2, the same for Room 13, two corner pieces of 2-in. thickness for each low partition, four miters for opening E, internal and external corners for the coved base, internal and external miters for the bullnosed cap at dado height.

There are no specials required for the ceramic glazed facing tile in Rooms 7 and 8.

Note that doorways are deducted at less than their full area where the deduction is from dado height facing tile. With a 12-course dado a single doorway (2–8 × 5–3) is deducted at 14 SF and double-door openings are deducted at 25 SF.

The take-off (W. D. 6.3, 6.4)

INTERIOR MASONRY

Clear glazed facing tile 5 × 12 2-in.

						OUTS				
93 ft	×	22c	=	2,046 pcs.	Drs.	2	×	18 SF	=	36 SF
99	×	12c	=	1,188		1	×	25	=	25
41	×	20c	=	820		4	×	14	=	56
				4,054	Opg.	6–0	×	7–0	=	42
		Less		724	Sash	21–0	×	4–2½	}	156
				3,330 pcs.		17–0	×	4–2½		
+	Waste 4%			130						315 SF
				3,460 pcs.	@ 2.3 per SF				=	724 pcs.

4-in. G1F

33 ft	×	22c	=	726 pcs.	Drs.	1	×	18 SF	=	18 SF
230	×	12c	=	2,760		1	×	35	=	35
41	×	20c	=	820		1	×	25	=	25
				4,306		5	×	14	=	70
		Less		414	Opg.	6–0	×	5–3	=	32
				3,892						180 SF
+	4%			158	@ 2.3 per SF				=	414 pcs.
				4,050 pcs.						

Chapter Six

4-in. G2F
```
    9 ft  ×  12c  =   108 pcs.
    8     ×  16c  =   128
                      ———
                      236
         +  4%          9
                      ———
                      245 pcs.
```

Extra for Group 1 =	180 pcs.	
2 =	625 pcs.	
3 =	320 pcs.	
4 =	78 pcs.	

Ceramic glazed facing tile 5 × 12 2-in.
```
                 26 ft  ×  19c     =  494 pcs.
    Less Drs.     2     ×  41 pcs. =   82
                                      ———
                                      412
                          +  4%        18
                                      ———
                                      430 pcs.
```

4-in. G1F
```
    26 ft  ×  19c  =  494 pcs.
          +  4%       21
                     ———
                     515 pcs.
```

4-in. G2F
```
    10 ft  ×  19c  =  190 pcs.
          +  4%       10
                     ———
                     200 pcs.
```

Wash down facing tile
```
                 3,330 pcs.
                 3,892
                   236
                   412
                   494
                   190
                 ———
                 8,554 pcs.
             =   3,720 SF
```

Truck facing tile
```
    2-in.   3,890  ×   7 lb  =  27,230 lb
    4-in.   5,010  ×  12     =  60,120
                                ———
                                87,350 lb
                             =    43.7 tons
```

8-in. concrete-block ptns.
```
    30-0  ×   1-6                      =   45 SF
    16-0  ×  11-2                      =  179
                                          ———
                                          224 SF
    Coursing 8 in. × 16 in. = 1⅛ per SF + ⅛  28
                                          ———
                                          252 pcs.
                              +  4%        10
                                          ———
                                          262 pcs. (8  ×  16)
```

6-in. concrete-block ptns.

30–0	×	8–8	=	290 SF
38–2	×	1–6	=	58
79–6	×	11–2	=	888
33–0	×.	6–0	=	198
60–0	×	2–5	=	145
				1,579 SF
	+	1/8		198
				1,777 pcs.
	+	4%		73
				1,850 pcs. (8 × 16)

OUTS

Drs.   2 × 18 = 36 SF

4-in. concrete-block ptns.

5–6	×	9–8	=	54 SF
21–0	×	1–6	=	33
84–4	×	6–0	=	506
48–0	×	11–2	=	536
36–0	×	2–4	=	84
26–0	×	3–0	=	78
19–0	×	8–9	=	166
41–4	×	5–3	=	217
				1,674
	−			126
				1,548 SF
	+	1/8		193
				1,741 pcs.
	+	4%		69
				1,810 pcs. (8 × 16)

Drs.   7 × 18 = 126 SF

3-in. concrete-block ptns.

53–8	×	4–4	=	232 SF
26–0	×	3–0	=	78
41–0	×	3–6	=	144
				454 SF
	+	1/8		57 pcs.
				511
	+	4%		19 pcs.
				530 pcs.

Notes on the take-off (W. D. 6.3, 6.4)

The facing-tile items are rounded off to even feet when transferred from the collection sheet. The deductions are changed from square feet to pieces before the total is transferred to the other side of the take-off sheet. The tile lays (12 in. long and $5\frac{1}{4}$ in. high) and equals 2.3 pieces per SF.

Four percent waste is allowed for all partition items. The waste for extra cost specials is added when transferring the items from the collection sheet.

For the 2-in. ceramic facing tile, the deduction is made in one step, using 41 pieces for a single door 18 SF times 2.3 pieces or 41 pieces.

For cleaning, the total quantity of all 2- and 4-in. tile is converted into square feet.

The item for trucking tile from the nearest railway siding to the job is figured by totaling the weight of the 2-in. material (allowing 7 lb per piece) and adding the 4-in. material (allowing 12 lb per piece). The weight of glazed facing tile varies, but can be obtained from the supplier.

The concrete block is all plastered and can be figured to course 8 in. (the most common coursing), which would require 9 blocks to cover 8 SF, or $1\frac{1}{8}$ blocks per SF.

MASONRY—MISCELLANEOUS ITEMS

Interior scaffolding, if required, is taken off in square feet. This item is the gross area of the interior partitions. The item should be taken off after the masonry take-off has been completed.

Mortar is taken off in cubic yards, more detailed information can be found on p. 121.

Brick anchors and ties are listed per thousand (M) and described.

Metal reinforcing for block partitions is taken off in lineal feet of specific width. The spacing requirements are usually specified. If not, the wall reinforcing should be figured for every second course.

Angle-iron lintels set by masons are taken off in the number of pieces.

Arch bricks are taken off in pieces, either separately from the rest of the brick or as an additional-cost item. Usually flat splayed arches require splayed bricks, which must be specially made to a pattern or cut on a brick saw at the jobsite.

Face brick laid in special patterns should be taken off as an extra-cost item, measured in square feet. For example, "Extra cost for face brick laid herringbone—...SF."

Molded brick courses should be taken off in pieces, either as a separate brick item or an extra-cost item.

Norman bricks are $11\frac{1}{2}$ in. $\times$ $2\frac{1}{4}$ in., and usually lay $11\frac{7}{8}$ in. brick plus joint: the vertical coursing would be as shown on the drawings.

Oversize bricks are any bricks that are greater in height than the usual $2\frac{1}{4}$ in. to $2\frac{3}{8}$ in. Usually, oversize face bricks will course three oversize bricks to four standard size.

Fire-clay flue lining is measured per lineal foot of a specific size.

Glass blocks are measured in pieces of a specific size (6 in. × 6 in., 8 in. × 8 in., 12 in. × 12 in.). Expansion material and reinforcing are usually included in the quotation for the glass blocks; thus those items are normally not taken off.

Mortar is an item that costs considerably more than many people realize. The theoretical quantity of mortar required to lay a given amount of masonry material is not difficult to compute, but the actual amount of mortar used will be much greater than that theoretical quantity. Standard face brick with ³⁄₈-in. joints requires (theoretically) 9 to 10 CF of mortar per 1,000 bricks, but on the job that would not be nearly enough. "Where does the mortar go?" is a question often asked. The answer is that some of it stays in the mixer, some spills out of the barrow, some sticks to the barrow, some sticks to the mortar board, some falls off the trowel and some is left at the end of the day's work. In addition, a few cement and lime bags are sure to burst and finally, much of the extra mortar goes into the wall for parging the brick batts or cut bricks. Regardless, it should be clear that theoretical quantities are not reliable guides for the estimator. The quantities suggested below are the results of many years of checking mortar quantities used on major jobs, yet for this problematical item, they must be regarded as approximate only.

Masonry material				*Mortar*	
Standard face brick or common brick	¼-in. joints			16 CF per	1,000 pcs.
Standard face brick or common brick	⅜-in. joints			18 CF per	1,000 pcs.
Standard face brick or common brick	½-in. joints			20 CF per	1,000 pcs.
Concrete blocks	8 in. ×	16 in.	4 in.	5 CF per	100 pcs.
Concrete blocks	8 in. ×	16 in.	6 in.	6 CF per	100 pcs.
Concrete blocks	8 in. ×	16 in.	8 in.	8 CF per	100 pcs.
Concrete blocks	8 in. ×	16 in.	10 in.	9 CF per	100 pcs.
Concrete blocks	8 in. ×	16 in.	12 in.	10 CF per	100 pcs.
Clay tile partition blocks	12 in. ×	12 in.	4 in.	4 CF per	100 pcs.
Clay tile partition blocks	12 in. ×	12 in.	6 in.	5 CF per	100 pcs.
Clay tile partition blocks	12 in. ×	12 in.	8 in.	6 CF per	100 pcs.
Clay tile partition blocks	12 in. ×	12 in.	12 in.	7 CF per	100 pcs.
Glazed facing tile			2 in.	4 CF per	100 SF
Glazed facing tile			4 in.	6 CF per	100 SF

Gypsum tile (gypsum mortar) tile	3 in.	3 CF per	100 SF
Gypsum tile (gypsum mortar) tile	4 in.	4 CF per	100 SF
Gypsum tile (gypsum mortar) tile	6 in.	5 CF per	100 SF
Fire brick — thin joint	fire clay	350 lb	per 1,000 pcs.
Fire brick — ordinary work	fire clay	550 lb	per 1,000 pcs.

Computing the total quantity of mortar (cubic yards) and transferring it as a separate item in the estimate is a much better practice than adjusting the unit prices of masonry items to allow for the mortar. This "juggling" method is very hit and miss, being perhaps an addition of $8 to $10 per M for the face brick and 5 to 10¢ for each backup block. Some examples of mortar mixes and costs are given on p. 000.

STONEWORK

Stonework may be taken off in various ways, depending on the most satisfactory unit of measurement for pricing. The standard unit of measurement used by cut-stone suppliers is cubic feet. The total quantity of stone shown on the bid from the stone company will often be considerably greater than the total on your estimate. The difference does not necessarily indicate that one of the two quantities is incorrect, because stone companies usually take off the items including waste.

Stonework features for exterior face-brick walls (such as sills, plinth courses, band courses, and trim around openings) should be taken off on the face-brick collection sheet. The stone quantity would be included in the "outs" from the face-brick item on the take-off sheet. Deductions for stonework will usually be from the face-brick item only; thus the backup area is the net face-brick area plus the "outs" for stonework.

If any appreciable quantity of stonework is involved, the requirements for stone anchors, dowels, cramps, etc., must be examined and taken off. Stone anchors and dowels are often specified in materials and sizes that sound simple enough but may prove very expensive. Bronze, brass, or galvanized dowels can be expensive. Two actual examples can be cited: a small job with $4,000 worth of polished marble exterior features where the stone anchors and dowels cost about $200 and a major seven-story building with granite facing where the cost of granite was $240,000 and the cost of stone dowels, cramps, and anchors was $3,500.

Boxing up and protecting stonework may be necessary or even specified; if so, an item should be taken off for it (usually as a lump sum). This item is frequently carried as a "protection" item under carpentry.

Working Drawing 6.5 shows a granite base course and limestone features for the building shown in W. D. 6.2. In taking off the stonework, W. D. 6.2 is used only for the plan layout and the window schedules.

The take-off (W. D. 6.2, 6.5)

STONEWORK

4-in. granite base course

$$883\text{--}4 \times 2\text{--}9 = \underline{2{,}430 \text{ SF}}$$

9-in. $\times$ 5½-in. limestone sills

A	4	$\times$	18–8	=	75 ft
E	18	$\times$	11–0	=	198
B, C, D	132	$\times$	4–9	=	627
					$\underline{900 \text{ LF}}$

3¾-in. 8¼-in. limestone band = $\underline{542 \text{ LF}}$

15-in. $\times$ 6-in. limestone coping = $\underline{1{,}102 \text{ LF}}$

Wash down stonework

Granite				=	2,430 SF
Sills	900–0	$\times$	1–3	=	1,125
Band	542–0	$\times$	0–9	=	407
Coping	1,102–0	$\times$	2–6	=	2,755
					$\underline{6{,}717 \text{ SF}}$

Box up & protect stone sills = $\underline{900 \text{ LF}}$

Truck stone from siding to job

$$2{,}155 \text{ CF} \times 180 \text{ lb} = \underline{194 \text{ tons}}$$

⅝-in. $\times$ 3-in. galv. dowels

Coping 293 $\times$ 2 = $\underline{586 \text{ pcs.}}$

Galv. strap anchors (twisted)

Band 140 $\times$ 1 = $\underline{140 \text{ pcs.}}$

Galv. dovetail anchors (split)

Base course 265 $\times$ 2 = $\underline{530 \text{ pcs.}}$

(continues on next page)

Nonstain cement mortar

$$2{,}155 \text{ CF conc.} \times 7 \text{ CF per } 100 \text{ CF conc.} = \underline{\underline{151 \text{ CF}}}$$
$$= \underline{\underline{6 \text{ CY}}}$$

Notes on the take-off (W. D. 6.2, 6.5)

The perimeter of the granite base course is the same as that of the face brick at the first floor. The limestone sills are taken off from the window schedule in W. D. 6.2, except that the stone should be 6 in. longer than the window opening. The band course item is the fourth-floor face-brick perimeter. The coping item is the sum of all the perimeters at roof level and is made up of:

Roof at 3rd floor	95–2 + 39–10 + 39–10	= 174–10
Roof at 4th floor	2 (93–8 + 37–6 + 37–6 + 25–6)	= 388– 4
Main roof	2 (204–4 + 65–2)	= 539– 0
		1,102– 2
		= 1,102 LF

The cleaning area includes all the exposed stone surfaces plus the sills (allowing for approximately two faces), the band course (measured to the nearest inch), and the coping (the sectional perimeter).

Protection will be needed only for the window sills. The item for trucking from the railway siding would be required only if the stone was shipped by rail. The items are reduced to cubic feet and the weight computed at 180 lb per CF (a figure that is only approximate and can be replaced by a more accurate figure if the actual density of the stone is known).

The dowels and anchors can be carried lump sum or by the unit. In taking off those items, one extra stone must be allowed for every straight run. For example, consider the dowels for the copings: 1,102 LF of coping in 4-ft lengths is 278 stones; but there are 15 straight runs (3 at the third floor, 8 at the fourth, and 4 at the main roof), totaling 293 stones requiring 2 dowels.

The nonstain cement mortar item allows 7 CF of mortar per 100 CF of stone. This is a reasonable average allowance for stonework that is in large pieces.

Stonework in general is taken off as follows:

Veneer or facing stone up to 8 in. thick in square feet of a stated thickness; 4- and 8-in. "in-and-out" courses in square feet.

Trim and feature panels over 12 in. in the smallest face dimensions in square feet of a stated thickness.

Trim, sills, jambs, etc., in lineal feet of a stated width and thickness.

Steps, buttresses, and entrance walls in cubic feet.

Special features (such as round columns or corner stones) in pieces, fully described (for example, "Circular fluted limestone cols., 20-in. dia. × 11–4 = 8 pcs.").

Flagstone paving in square feet of a stated thickness.

Rubblestone walls in cubic feet.

Chapter 7

Carpentry

The specifications must be checked for any special requirements noted on the take-off sheets. Framing lumber is usually dressed on all four sides, and is only nominal of the given dimensions. For example, 2 in. × 4 in. D4S is actually $1\frac{1}{2}$ in. × $3\frac{1}{2}$ in., and a 1-in. board is $\frac{3}{4}$ in. when dressed. Should the specifications call for finish sizes, care must be taken to check those sizes against the standard and the special dimension. As an example, a job that calls for roof boarding of $\frac{7}{8}$-in. finish (D2S), the material cannot be obtained from a 1-in. rough board. It will require $1\frac{1}{4}$-in. material. Thus, the material price should be checked with the supplier and priced accordingly.

The preservative treatment required for wood should be carefully checked and the requirements noted on the take-off sheet. Pressure treatment of lumber is expensive and more costly than the lumber itself. Material costing $120 per 1,000 BF originally might cost another $160 per 1,000 BF for pressure preservative treatment. Further, should the job be in an area that contains no facilities for pressure treatment of lumber, the material may need to be trucked to the pressure tanks and back, all of which must be included as part of the cost of the item. A specification calling only for "wood preservative treatment," however, is a different matter, a simple dipping or a brush coat could serve.

The take-off examples cover (1) simple blocking and furring for a masonry building with wood doors and double-hung wood windows, (2) typical classroom requirements for a school, and (3) the majority of the carpentry for a wood framed two-family house.

EXTERIOR CARPENTRY

Working Drawing 7.1 shows the exterior wall details for the building in W. D. 6.2. The carpentry take-off includes all the rough and finish items. Additional information is taken from the specifications: Exterior wood blocking at roof, eaves nailers and cants to be pressure treated "wolmanized," rough bucks for doors are to have three galvanized anchors $\frac{1}{8}$ in. × $1\frac{1}{4}$ in. × 8 in. per jamb. Exterior doors are to be 3 ft × 7 ft, 2-in. solid pine, 6 panel, the pair of doors has 2-in. molded and beveled astragals (one for each leaf). Ceiling height to vary from 8 ft 6 in. to 9 ft 4 in., all exterior trim is to be clear pine and all interior is trim birch. The ends of window stools are notched and returned, window frames and sashes are white pine, all doors, door frames, window frames, and sashes are primed at the mill, all other millwork and trim is back-primed before erection. The total number of rooms is 112.

The take-off (W. D. 6.2, 7.1)

ROUGH CARPENTRY

3 × 12 eaves nailers (pressure wolmanized)

3rd	96–0 + 42–0 + 42–0		=	180 ft	
4th	2(94–0 + 40–0 + 40–0 + 28–0)		=	404	
Main	2(206–0 + 68–0)		=	548	
				1,132 LF	
			=	3,396 BF	
			=	3,400 BF	

1 × 6 eaves nailer (pres. wolm.) = $\underline{1,140 \text{ LF}}$

4-in. roof cants (pres. wolm.)

$$96-0 + 68-0 + 68-0 = \underline{232 \text{ LF}}$$

½-in. × 12-in. bolts for conc.

$$
\begin{aligned}
1{,}132 \text{ at } 1 \text{ per } 3 \text{ ft} &= 378 \text{ pcs.} \\
+\ 1 \text{ per run} &= \underline{15} \\
&= 393 \text{ pcs.} \\
\text{Say,} &\ \underline{400} \text{ pcs.}
\end{aligned}
$$

Window blocking

Head	2 × 6	4/20 +	18/12 +	132/ 5	=	956 ft	}	2,468 BF	
Jambs	2 × 6	84/ 6 +	112/ 5 +	112/ 4	=	1,512	}		
Sill	2 × 4	16/12 +	64/ 5		=	512	=	342	
Sill	2 × 8	4/20 +	68/ 5 +	2/12	=	444	=	592	
								3,402 BF	
							=	3,400 BF	

1 × 4 blocking at mullions

	A	4 × 3 × 6–0	}	288 ft			
	E	18 × 2 × 6–0	}				
			=	290 LF			

1 × 2 wall furring for plaster

1st & 2nd	2(877	+	12)	=	1,778 pcs.			
3rd	795	+	8	=	803			
4th	535	+	4	=	539			
					3,120 pcs.	× 10 ft	=	31,200 LF
	Add 1 length per room			112		× 10	=	1,120
								32,320 LF

¾-in. × 4-in. ground for metal base

$$880{-}0 \;+\; 880{-}0 \;+\; 800{-}0 \;+\; 540{-}0 \;=\; \underline{3{,}100 \text{ LF}}$$

2 × 8 rough bucks for door frames

$$3 \times 24{-}0 = 72 \text{ ft}$$
$$7 \times 20{-}0 = 140$$
$$\overline{212 \text{ LF}}$$
$$= \underline{284 \text{ BF}}$$

⅛ × 1¼ × 8 galv. anchors

$$10 \times 2 \times 3 = \underline{60 \text{ pcs.}}$$

MILLWORK (SET ONLY)

Wood window frames (white pine, primed)

$$18{-}2 \times 5{-}4 = \underline{4 \text{ fr.}}$$
$$10{-}6 \times 5{-}4 = \underline{18 \text{ fr.}}$$
$$4{-}3 \times 5{-}4 = \underline{20 \text{ fr.}}$$
$$4{-}3 \times 4{-}7 = \underline{56 \text{ fr.}}$$
$$4{-}3 \times 3{-}8 = \underline{56 \text{ fr.}}$$

D.H. sashes (spiral balances)

$$4 \times 3 = 12$$
$$18 \times 2 = 36$$
$$132$$
$$\overline{180 \text{ pairs}}$$

Ext. wood door frame 2 in. × 9 in., 3 ft × 7 ft = $\underline{7 \text{ fr.}}$

Ext. wood door frame 2 in. × 9 in., 6 ft × 7 ft = $\underline{3 \text{ fr.}}$

2-in. pine ext. doors 3 ft × 7 ft = $\underline{13 \text{ dr.}}$

2-in. astragal 7 ft long = 6 pcs.

Clear pine staff bead 3 in. × *2 in. (back-primed)* = 2,468 LF

Clear pine mullion trim 1 × *3½ (back-primed)* = 290 LF

EXTERIOR CARPENTRY

Birch window trim (back-primed) 1 × *4 cover mold* = 2,468 LF

Birch window trim (back-primed) 1-in. scotia = 2,468 LF

Birch window trim (back-primed) 1 × *8 stool (notch & return)*

$$\begin{array}{rcl} 4 \times 20\text{-}0 & = & 80 \text{ ft} \\ 68 \times 5\text{-}0 & = & 340 \\ 2 \times 12\text{-}0 & = & 24 \\ \hline & & 444 \text{ LF} \end{array}$$

Birch window trim (back-primed) 1 × *12 stool (notch & return)*

$$\begin{array}{rcl} 16 \times 12\text{-}0 & = & 192 \text{ ft} \\ 64 \times 5\text{-}0 & = & 320 \\ \hline & & 512 \text{ LF} \end{array}$$

Birch window trim (back-primed) ¾ × *4 apron* = 956 LF

Birch window trim (back-primed) 1 × *2½ mullion trim* = 290 LF

1 × *4 door casing (birch, back-primed)* = 284 LF

Set finish hardware = L. S.

Rough hardware = L. S.

Temporary doors = 10 opgs.

Notes on the take-off (W. D. 6.2, 7.1)

The eave nailer item reflects the roof dimensions to the nearest even-foot multiple. The front roof at the third floor is actually 95 ft 2 in. × 40 ft 10 in., for which we take off 96 ft plus 42 ft plus 42 ft. Continuous nailers in long runs do not require waste allowance unless the specifications require the joints to be half-lapped. For all items such as eave nailers or sill plates requiring lapped joints, an allowance should be made of 6 in. for every 14 ft, and the total quantity increased accordingly. If the eave blocking had been specified to require lapped joints, the total length in lineal feet would have been 1,132 LF plus 42 LF, or 1,174 LF. The total quantity is converted from lineal feet to board feet. One BF of lumber is 1 LF times 12 sq. in. in section. Thus a 3×12 is 3 BF per LF. Board feet of lumber can be calculated by multiplying the lumber thickness in nominal inches $\times$ length $\times$ width divided by 144.

The 1×6 nailer has the same length as the 3×12, but is rounded from 1,132 LF to 1,140 LF. Notice that the 1×6 is not converted to board feet. For members not more than $\frac{1}{2}$ BF per LF, the lineal foot measure may be better for pricing, since small items are usually more costly to handle. These should be left in lineal feet rather than lumped in with the larger items in board feet.

There should be an extra bolt added for the end of each run of wall. Many estimators do not take off anchor bolts. The take-off is simple and should be added to the estimate.

The window blocking follows the window schedule, with even-foot multiples taken off for everything over 7 ft and odd-foot multiples for lengths under 7 ft. The 2×6 at the head is for the A-type windows: 4 windows at 18 ft 2 in. long require four pieces at twenty foot of 2×6 blocking. Notice that the sill blocking is 2×4 for the third and fourth floors, but 2×8 for the first and second floors. The two sill totals are the same as the head item (512 ft plus 444 ft = 956 ft).

The 1×4 blocking at the mullions may not be needed if the window frames are completely mill-assembled. It would be required if the big windows were to be delivered with the mullion material bundled for job assembly.

To determine the required amount of the 1×2 furring, take off the perimeter at each floor and add one (10-ft) piece for each run of wall plus an additional piece for every room. The first floor inside perimeter is 876 ft 8 in., rounded to 877 ft then add twelve 10-ft pieces for the wall runs for each corner. This gives 889 pieces of furring (1 piece per LF) for each of the first two floors for 1,778 pieces total for two floors. Repeating this method for the third and fourth floors yields a total of 3,120 pieces for 31,200 LF. The furring is taken off at 10 ft per piece because the ceiling heights vary from 8 ft 6 in. to 9 ft 4 in. Therefore, 10-ft lengths are required. There are 112 rooms in the building. Each room will require an extra piece at the corners. Therefore, we add 112 additional pieces. From the plan layout, one would count the rooms involved to determine this allowance. It should be noted that no allowance has been made for window openings. Unless the openings are either predominantly very large or very small, it is not necessary to allow for them in the wall furring item. It is usually found that the furring required around the perimeter of the openings will be offset by the straight-run vertical material. Calculating the window openings in this example will show that the adjustment for windows would yield an additional 40 LF of furring, which is not significant in a total of 32,000 LF. This calculation is worth considering in detail:

18–2 × 5–4 window

Deduct	17 × 6–0					
Add	2 × 18–0	} Deduct	56–0 × 4	=	224 ft
Add	2 × 5–0					

10–6 × 5–4 window

Deduct	9 × 6–0	} Deduct	22–0 × 18	=	396
Add	2 × 16–0				
				−	620

4–3 × 5–0 average window

Deduct	3 × 5–0	} Add	5–0 × 132	=	660
Add	4 × 5–0				
			Net	+	40 LF

If the wall furring were spaced on 14- or 16-in. centers, additional furring will probably be required for the windows. In this case, the straight runs would be less than required around the openings.

The grounds for the metal base is the sum of the floor perimeters rounded off. The other rough carpentry items are self-explanatory. It should be noted at the end of the take-off, following the millwork, there are two items, rough hardware and temporary doors, that belong with the rough carpentry. Those items should be entered on the estimate sheet with the rough carpentry. These were missed during the original take-off.

The wood doors, windows, and millwork have been taken off for setting only. This material will be supplied by a subcontractor. In taking off millwork for setting only, it is neither necessary to describe in great detail nor to break down by sizes. The cost to hang sashes and doors according to sizes will not vary. Therefore, no advantages are gained by breaking down these units by size. Triple and double window frames should be separated from the single ones. Likewise, very large doors ($3\frac{1}{2}$ ft wide and over) should be separated from the more usual 3-ft and under sizes.

The take-off of wood window frames, sashes, and doors is straightforward. All the single window frames should be priced at the same unit cost for setting.

The window trim is taken off after the blocking. The staff bead length is the same as that of the 2 in. × 6 in. blocking at heads and jambs. Referring to the blocking take-off, we have 956 LF plus 1,512 LF (2,468 LF) of staff bead. All window trim quantities come from the blocking take-off.

The finish hardware is usually scheduled in the specifications and can be taken off directly from there. Very often the finish hardware is an allowance item; consequently, the setting labor may be priced as a lump-sum percentage of the allowance amount. A better method would be to price the finish hardware with the door installation.

A CLASSROOM

Working Drawings 7.2 and 7.3 show the plan and details for a typical classroom in a school that is to have 12 similar rooms. The take-off will include all rough carpentry; setting of all millwork and casework; setting doors and frames; and material labor for chalkboard, tackboard, resilient flooring, and base.

The take-off (W. D. 7.2, 7.3)

1 × 2 wall furring

$$
\begin{array}{rcl}
16 \times 7\text{–}0 &=& 112 \text{ ft} \\
13 \times 4\text{–}0 &=& 52 \\
4 \times 4\text{–}0 &=& 16 \\
10 \times 6\text{–}0 &=& \underline{60} \\
& & 240 \text{ ft} \times 12 \\
&=& \underline{\underline{2{,}880 \text{ LF}}}
\end{array}
$$

1 × 3 grounds for casework

Teacher's closet	3 ×	10–0	=	30 ft
Sink cab.	2 ×	6–0	=	12
Window cab.	2 ×	30–0	=	60

$$
\begin{array}{rcl}
& & 102 \text{ ft} \times 12 \\
&=& \underline{\underline{1{,}224 \text{ LF}}}
\end{array}
$$

Blocking for cabinets

$$
\begin{array}{l}
2 \times 4 \quad \left. \begin{array}{l} 2 \times 6\text{–}0 \\ 2 \times 30\text{–}0 \end{array} \right\} \quad 72 \text{ ft} \\
\hspace{4cm} \times 12 \\
\hspace{3.5cm} \overline{864 \text{ LF}} \\
\hspace{2.5cm} = \quad \underline{\underline{580 \text{ BF}}}
\end{array}
$$

⅜-in. plywood behind chalk and tack bds.

$$
\left.\begin{array}{l} 12 \times 12\text{–}0 \times 6\text{–}0 \\ 12 \times 36\text{–}0 \times 4\text{–}6 \end{array}\right\} = 12 \times 234 \text{ SF} = 2{,}808 \text{ SF}
$$

$$
\begin{array}{rl}
+ \text{ Waste } 10\% & \underline{282} \\
& \underline{3{,}090 \text{ SF}}
\end{array}
$$

⅜-in. plywood G1S to wall dado

$$
\begin{array}{rrcr}
& 12 \times 19\text{–}6 \times 6\text{–}6 & = & 1{,}521 \text{ SF} \\
\text{Less} & 12 \times 16\text{–}0 \times 4\text{–}6 & = & \underline{864} \\
& & & 657 \\
& + \ 10\% & & \underline{63} \\
& & & \underline{720 \text{ SF}}
\end{array}
$$

¼-in. cork tackboards

$$
\begin{array}{rcrcrcrcl}
 & & 12 & \times & 12\text{-}0 & \times & 1\text{-}4 & = & 192 \text{ SF} \\
12 & \times & 3 & \times & 4\text{-}0 & \times & 4\text{-}0 & = & \underline{576} \\
 & & & & & & & & 768 \\
 & & & & & & +\ 5\% & & \underline{38} \\
 & & & & & & & & \underline{806 \text{ SF}}
\end{array}
$$

⅛-in. porc. steel chalkboards (cemented to ply)

$$
\begin{array}{rcrcrcl}
12 & \times & 36\text{-}0 & \times & 4\text{-}0 & = & 1{,}728 \text{ SF} \\
 & & & & +\ 5\% & & \underline{82} \\
 & & & & & & \underline{1{,}810 \text{ SF}}
\end{array}
$$

Rough hardware = <u>L. S.</u>

MILLWORK, ETC., TO SET

Int. H. M. door frames = <u>24 ea.</u>

Int. H. M. door & hardware = <u>24 ea.</u>

Teacher's closet 2–6 × 2–0 × 9–0 = <u>12 ea.</u>

Sink counter 6–0 × 2–0 × 2–4 = <u>12 ea.</u>

Window counter 30–0 × 2–0 × 2–4 = <u>12 ea.</u>

Hang cabinet doors (2 ft × 2 ft) & hardware

$$12 \times 14 = \underline{168 \text{ drs.}}$$

1 × 6 *backboard over counters*

$$12 \times 36 \text{ ft} = \underline{432 \text{ LF}}$$

1 × 4 *chalk tray*

$$12 \times 36 \text{ ft} = \underline{432 \text{ LF}}$$

1 × 4 *trim at C.B. & T.B.*

$$
\left.\begin{array}{rcr}
12 & \times & 52 \text{ ft} \\
12 & \times & 20 \\
12 & \times & 56 \\
12 & \times & 46
\end{array}\right\} 12 \times 174 \text{ ft}
$$

$$= \underline{2{,}088 \text{ LF}}$$

$\tfrac{3}{8} \times \tfrac{1}{2}$ trim at C.B. & T.B.

$$
\left.\begin{array}{r} 12 \times 60 \text{ ft} \\ 12 \times 16 \\ 12 \times 48 \\ 12 \times 40 \end{array}\right\} \quad 12 \times 164 \text{ ft}
$$
$$= \underline{1{,}968 \text{ LF}}$$

$\tfrac{3}{4} \times 2$ trim at C.B. & T.B.

$$
\left.\begin{array}{r} 12 \times 52 \text{ ft} \\ 12 \times 20 \\ 12 \times 56 \\ 12 \times 20 \end{array}\right\} \quad 12 \times 148 \text{ ft}
$$
$$= \underline{1{,}776 \text{ LF}}$$

$\tfrac{1}{8}$-in. lino. to counter tops

$$12 \times 36\text{–}0 \times 2\text{–}0 = \underline{864 \text{ SF}} \text{ (net)}$$

$\tfrac{1}{8}$-in. lino. to 6-in. backboard $= \underline{432 \text{ LF}}$ (net)

4-in. rubber tile coved base

$$12 \times 102 \text{ ft} = \underline{1{,}224 \text{ LF}} \text{ (net)}$$

9-in. × 9-in. × $\tfrac{1}{8}$-in. asphalt-tile floor (2 colors)

$$12 \times 30\text{–}0 \times 24\text{–}0 = \underline{8{,}640 \text{ SF}}$$

Notes on the take-off (W. D. 7.2, 7.3)

Wall furring for the 19 ft. 6 in. wall, has plywood 6 ft 6 in. high that requires 15 pieces of 1 × 2 at 16-in. centers plus one for the end, for a total of 16 pieces at 7 ft. The other furring items are for the 16-ft combination of chalkboard and tackboard, the 4-ft tackboard, and the 12-ft chalkboard-tackboard combination. The furring length for one room is multiplied by 12.

The grounds and blocking for the casework are straightforward items. The plywood backing is taken off first at elevation A for the 4-ft. chalkboard and 1 ft 4 in. tackboard above. The plywood will be 6 ft high as shown on the section. The other boards are all 4 ft high with 4 ft 6 in. plywood backing. Consequently, these are combined into one item, 4 ft plus 16 ft plus 16 ft (36 LF). A waste allowance of 10 percent is added for cutting.

The exposed plywood is separated from the backing plywood because they are different materials. The dado plywood must have one good face for a paint finish requiring greater care in setting.

The tackboard and chalkboard are taken off at the full 4-ft sheet sizes. The three 1 ft 4 in. items will cut a 4-ft sheet. Only a 5 percent waste allowance is needed for these items, because there is no waste in the width of the material.

The casework and trim items are straightforward. The trim is measured in even feet per board (chalk and tack). The board in elevation A requires mitered trim around the board. A 12-ft piece would be too short to cover a 12-ft board; therefore, the top and bottom trim must be taken off at 14 ft. The 12-ft length will do for the intermediate horizontal piece. The sides are 6 ft each. This gives a total of 14 ft plus 12 ft plus 14 ft plus 6 ft plus 6 ft or 52 LF per board. Notice that although 14 ft is taken off a net 12-ft board in the length, the height is taken off at the net 6 ft and not 7 ft. This is because it is possible to cut one miter without waste. In order to cut the opposite miter, however, a longer length will be required.

The resilient flooring, base, and countertops may be measured net as long as it is clearly noted. The vinyl tile floor is measured to the face of the continuous window cabinet (26 ft less 2 ft). The small cabinet and closet are not deducted.

WOOD-FRAMED BUILDINGS

Working Drawings 7.4 and 7.9 show a wood-framed duplex house, as may be found in private home construction or government housing projects. If you can take off a single building accurately, with little additional effort, you can take off an entire project of 2,000 dwelling units. In taking off a large housing job that consists of many identical buildings, it is acceptable to take off one building, price it, and multiply the total by the number of buildings. This procedure will usually apply only to the work above ground, because the foundation depth and excavation required may vary from building to building. The multiplication procedure becomes somewhat complex and must be handled carefully with all the steps clearly shown on the estimate sheets.

A housing project will include several types of buildings. It is best to take off the entire job progressively. Each item should be taken off by measuring the quantity for one type of building and multiplying the results by the number of buildings of that type. Then repeat the procedure for each building type. This method has its advantages. It reduces the paperwork in the take-off, provides the actual total quantities for each item, and gives the actual purchase quantities.

Combining repetitive items or items of similar design can be accomplished in a housing project with similar types of buildings. All of the information can be summarized on a single spreadsheet that will greatly simplify the take-off. A typical collection of several items is shown in Table 7.1 for an actual job of 250 dwelling units.

Combining the information shown in Table 7.1 not only saves hours of estimating time but also simplifies the subsequent calculations and decreases the probability of errors considerably. The take-off for this $2,800,000 project was the same as shown for the single building in the example that follows.

Working Drawings 7.4 to 7.9, showing a two-story duplex dwelling, are taken off in two parts; first, the rough carpentry framing (walls, partitions, roof framing, boarding, and exterior wall finish siding, roof shingles, and gut-

TABLE 7.1 Collection Sheet for Wood-Framed Buildings

Bldg. type	Dwelling units			Perimeter		Area		Fire walls		Roof
	Bldgs.	D.	U.	1 bldg.	Total	1 bldg.	Total	1st	2nd	
A	25 × 4	=	100	179–4	4,483 ft	1,503 SF	37,575 SF	550 ft	600 ft	600 ft
B	24 × 4	=	96	210–0	5,040	1,904	45,696	541	552	552
C	12 × 4	=	48	199–8	2,396	1,963	23,556	308	308	308
D	1 × 6	=	6	292–2	292	2,886	2,886	45	45	45
			250		12,211 ft		109,713 SF	1,444 ft	1,505 ft	1,505 ft

Bldg. type	Bast. sash		Entrance doors		Windows			
	1	2	A	B	1	2	3	4
A	25	25	100	100	400	400	300	400
B	24	24	96	96	576	480	288	384
C	18	6	36	60	288	192	192	192
D	2	1	—	12	24	24	12	24
	69	56	232	268	1,288	1,096	792	1,000

ters), and second, the interior furring, grounds, blocking, plywood, stairs, millwork, and finish carpentry.

The exterior

The following items and description of materials supplement the drawings: wall studding is to be 2 × 4 on 16-in. centers unless otherwise indicated (sill bolts have been previously taken off). Sills and girts are to be lapped at joints and corners. Fire stops, the same depth as joists, are required at first- and second-floor exterior walls. All boarding is to be 1 in. × 8 in. (laid diagonally for walls and subfloors). The exterior wall finish is to be ¾ in. × 8 in. bevel siding, 6¾ in. exposed with 15-lb felt underlayment. Asphalt shingles for roof are to be 3-tab, 12 in. × 36 in., 5 in. exposed to weather; cedar shingles are to be used in starter course at eaves and 30-lb roof underlayment. Soundproof partitions are to be two 2 × 4 stud partitions with offset studs and 4-in. blanket insulation woven between the studs. The attic outer ceiling space is to have 2-in. blanket insulation with vapor-barrier paper on the underside. Include 15-lb felt over all subflooring. The gutters are to be 4 in. × 3 in. redwood, primed twice with half-lapped joints and lead inserts. The Lally columns supporting the beam at the first floor may be considered as having been previously taken off.

The small checkmarks on the drawings were made as the items were being taken off.

The take-off (exterior—W. D. 7.4–7.9)

4 × 6 sill

$$2\,(50\text{ ft} + 24\text{ ft}) = 148\text{ ft}$$
$$= \underline{300\text{ BF}}$$

Floor frmg.

1st 2 × 8 (half of floor)
$$23 \times 16\text{-}0 = 368\text{ ft}$$
$$24 \times 10\text{-}0 = 240$$
$$4 \times 4\text{-}0 = \underline{16}$$
$$\overline{\overline{624\text{ ft}}} \times 4 = 2{,}496\text{ ft}$$
$$= \underline{3{,}328\text{ BF}}$$

4 × 6 beam

$$50\text{ ft} = \underline{100\text{ BF}}$$

1 × 3 cross bridging

$$4 \times 42 \times 2\text{-}6 = \underline{420\text{ LF}}$$

4 × 4 posts

$$2 \times 3 \times 10\text{-}0 = \underline{80\text{ BF}}$$

Joist hangers (for 2 × 8) = $\underline{8\text{ pcs.}}$

Joist hangers (for double joist) = $\underline{8\text{ pcs.}}$

Ceiling joists 2 × 6

$$4 \times 19 \times 14\text{-}0 = \underline{1{,}064\text{ BF}}$$

Roof frmg.

Hips $\quad 2 \times 10\text{-}0 \quad 4 \times 20\text{-}0 = 80\text{ ft}$
Ridge $\quad 2 \times 10\text{-}0 \qquad\qquad\; 26\text{-}0 = \underline{26}$
$$\overline{\overline{106\text{ ft}}}$$
$$= \underline{180\text{ BF}}$$

Rafter 2 × 8

$$23 \times 14\text{-}0 = 322\text{ ft}$$
$$8 \times 8 \times 7\text{-}0 = \underline{448}$$
$$= \overline{\overline{770\text{ ft}}}$$
$$= \underline{1{,}030\text{ BF}}$$

Collar ties 2 × 6

$$21 \times 8\text{-}0 = \overline{\overline{168\text{ ft}}}$$
$$= \underline{170\text{ BF}}$$

Hangers 1 × 6

$$21 \times 2 \times 5\text{-}0 = \underline{210 \text{ LF}}$$

1 × 6 cont. bridging

$$2 \times 50\text{-}0 = \underline{100 \text{ LF}}$$

Wall frmg. 4 × 6 girt

$$148 \text{ ft} = \underline{300 \text{ BF}}$$

Wall frmg. 4 × 6 posts

$$
\begin{aligned}
7 \times 8\text{-}0 &= 56 \text{ ft} \\
7 \times 10\text{-}0 &= \underline{70} \\
&\ 126 \text{ ft} \\
&= \underline{252 \text{ BF}}
\end{aligned}
$$

Wall frmg. 2 × 4

Head		2 ×	148-0	=	296 ft
1st	2 ×	47 ×	8-0	=	752
2nd	2 ×	47 ×	10-0	=	940
Wind. V	2 ×	26 ×	4-0	=	208
Door	2 ×	4 ×	1-0	=	8
Door head	4 ×	2 ×	4-0	=	32
Wind. H & S	6 ×	4 ×	6-0	=	144
Wind. H & S	14 ×	4 ×	5-0	=	280
Wind. H & S	2 ×	4 ×	4-0	=	32
Bracing		8 ×	8-0	=	64
Bridging	2 ×	61 ×	1-3	=	154
					2,910 ft
				=	1,940 BF

2 × 8 firestopping

$$2 \times 2 \times 50\text{-}0 = \underline{270 \text{ BF}}$$

1 × 8 T. & G. subflooring

$$
\begin{aligned}
2 \times 47\text{-}6 \times 22\text{-}4 &= 2{,}122 \\
\text{Less} \quad 4 \times 9\text{-}0 \times 3\text{-}8 &= \underline{132} \\
&\ 1{,}990 \text{ SF} \\
+ \text{ Waste } 20\% &\ \underline{400} \\
&\ \underline{2{,}390 \text{ BF}}
\end{aligned}
$$

15-lb felt over subflooring

$$+\ 10\% \quad \frac{\begin{array}{r}1{,}990\ \text{SF}\\ 200\end{array}}{2{,}190\ \text{SF}}$$

1-in. × 8-in. roof bdg. (T. & G.)

$$2\ \times\ 48\text{-}2\ \times\ 14\text{-}0\ =\ 1{,}349\ \text{SF}$$
$$+\ \text{Waste}\ 15\% \quad \frac{201}{1{,}550\ \text{BF}}$$

30-lb roofing felt

$$+\ 10\% \quad \frac{\begin{array}{r}1{,}349\ \text{SF}\\ 135\end{array}}{1{,}484\ \text{SF}}$$
$$=\ \underline{1{,}500\ \text{SF}}$$

Cedar shingle starter course

$$142\text{-}4\ \text{at}\ 0\text{-}4\ =\ 427\ \text{pcs.}$$
$$+\ \text{Waste} \quad \frac{23}{450\ \text{pcs.}}\ (4\ \text{in.}\ \times\ 16\ \text{in.})$$

Asphalt shingles, 3 tab, 12 in. × 36 in., 5 in. to weather

$$1{,}350\ \text{SF}\ =\ 13.5\ \text{sqrs.}$$
$$+\ \text{Waste}\ 7\% \quad \frac{1.0}{14.5\ \text{sqrs.}}\ (\text{to cover})$$

4-in. × 3-in. redwood gutter (twice-primed)

$$2\ (52\ \text{ft}\ +\ 26\ \text{ft})\ =\ \underline{156\ \text{LF}}$$

Half-lapped joints + lead insert = $\underline{14}$ jnts.

2-in. blanket insulation over clg. = $\underline{2{,}200\ \text{SF}}$ (+ vapor barrier 1 side)

Alum. foil insulation to ext. walls.

	139-8	×	17-4			=	2,422 SF
Door	4	×	20 SF		=	80	
Sash A	4	×	6-0	×	4-0	=	96	
B	8	×	4-6	×	4-0	=	144	480
C	6	×	4-0	×	4-0	=	96	
D	2	×	3-0	×	4-0	=	24	
E	2	×	5-0	×	4-0	=	40	

$$+\ 10\% \quad \frac{\begin{array}{r}1{,}942\\ 198\end{array}}{2{,}140\ \text{SF}}$$

1 × 8 T. & G. wall sheathing

$$
\begin{array}{rrcl}
142\text{--}4 \times 18\text{--}4 & = & 2{,}610 \text{ SF} \\
\text{Less Opgs.} & = & \phantom{2{,}}480 \\
\hline
& & 2{,}130 \text{ SF} \\
+\ 20\% & & \phantom{2{,}}430 \\
\hline
& & \underline{2{,}560 \text{ BF}}
\end{array}
$$

15-lb felt to walls

$$
\begin{array}{rcl}
& & 2{,}130 \text{ SF} \\
+\ 10\% & & \phantom{2{,}}220 \\
\hline
& & \underline{2{,}350 \text{ SF}}
\end{array}
$$

¾-in. × 8-in. bevel siding (6¾ in. to weather)

$$
\begin{array}{rcl}
& & 2{,}130 \text{ SF} \\
+\ \text{Waste } 28\% & & \phantom{2{,}}600 \\
\hline
& & \underline{2{,}730 \text{ SF}}
\end{array}
$$

Int. stud ptns. 2 × 4

1st Sill and head	3	×	118–0			=	354 ft
Studs	112	×	8			=	896
+ Cupb.	2	×	20			=	40
2nd Sill and head	3	×	188			=	564
Studs	175	×	8			=	1,400
+ Cupb.			40			=	40
Low ptns.	2	×	2	×	6	=	24
Low ptns.	2	×	6	×	4	=	48
Bridging	310–0					=	310
Add for doors	10	×	16			=	160
							3,836 LF
						=	2,560 BF

4-in. blanket insulation to ptns.

$$
\begin{array}{llcll}
\text{1st} & 29\text{--}0 \times 8\text{--}2 & = & 237 \text{ SF} \\
\text{2nd} & 24\text{--}0 \times 8\text{--}0 & = & 192 \\
& & & \overline{429} \\
& +\ 15\% & & 66 \\
& & & \overline{495 \text{ SF}} \\
& & = & \underline{500 \text{ SF}}
\end{array}
$$

Notes on the take-off (exterior—W. D. 7.4–7.9)

The exterior sill item, taken off at 50-ft and 24-ft lengths, allows sufficient material for both laps and 2-ft multiples. The floor joists were taken from the framing plan by taking off half of one floor and then multiplying by 4 to obtain the total for the two floors. The cross bridging item allows two 2 ft 6 in. pieces for each joist spacing. Joist hangers are taken off for the framing at the stairwells.

The hip is shown on the roof plan (dotted line). This is accomplished by setting up the rise perpendicular to the hip in plan, then projecting the full dimensions of the hip onto the plan, appearing to be flipped over to rest on the roof. In this particular roof, the hipped ends are at the same pitch as the main roof line. Therefore, the hips form a right angle on the plan. The rise is laid out on one hip and the hip length is measured along the dotted line. Since it measures 17 ft 6 in. with a long cut at each end and 18-ft-length material is not available, the hip was taken off as 20-ft material.

The main rafters scale 13 ft 6 in., so a 14-ft length is used. The hip rafters must average half the main rafters (7 ft); therefore, the quantity is eight times 7 ft. The collar ties and hangers are scaled from the drawing.

The 4×6 girt is the same length as the 4×6 sill. The 4×6 posts are 8 ft for the first floor, and 8 ft 4 in. for the second (requiring 10-ft material).

The exterior 2×4 framing is added up from the wall-framing drawing. The under and over window studs are 4 ft to allow for cutting. The window sill and head plates are doubled 2×4's 6 ft long.

The 2×8 firestop runs along the front and rear walls across the open ends of the joists. This item is taken off at the full length of the building, 48 ft 2 in. and rounded up to 50 ft.

The subflooring runs to the inside face of the exterior studs. The nominal 8-in. tongue-and-groove board finishes to $7\frac{1}{4}$ in. or $9\frac{1}{2}$ percent loss. The waste factor from cutting and fitting typically is 20 percent. The felt underlayment is the same as the subfloor quantity.

A rule of thumb for roof area is: a 45° hipped roof has approximately the same area as a gable-end roof. Therefore, a quick calculation is twice the roof length times the slope length. Add 15 percent for the cedar-shingle roof board waste. The cedar-shingle starter course is measured in pieces, not in squares. The asphalt shingles are in squares of area to cover. In ordering asphalt shingles, the size and the exposure are given; therefore, one square will cover 100 SF. For example, one square of 3 tab 12 in. $\times$ 36 in. shingles, 5 in. to weather, would consist of 80 pieces (80 times 3–0 times 0–5 gives 100 SF). The same shingle with 4 in. to weather would have 100 pieces per square.

The redwood gutter is measured with an allowance for the half-lapped joints. The expensive joint with a lead insert is measured as a separate item. There are four joints along each of the front and back gutters, one joint on each end run, and four corner miters for a total of 14 joints.

The ceiling insulation item is the same area as the flooring. The wall insulation is measured inside the studs, the openings deducted, and the waste added. The area for the exterior wall sheathing is the exterior perimeter measured from the bottom of the sill to the top of the eaves (0–4 plus 0–8 plus 8–10 plus 8–6, a total of 18–4), openings deducted and 20 percent waste added, to allow for diagonal laying as specified. The felt area is equal to the sheathing area plus waste. The bevel siding measures $7\frac{1}{4}$ in. but will lay $6\frac{3}{4}$ in., a loss of $1\frac{1}{4}$ in. per 8-in. board or 16 percent. Allowing an additional 12 percent for cutting and fitting results in a total waste of 28 percent.

The stud partitions are measured from the plan (in even-foot lengths). Taking off the first floor: there are 60 LF of partition (14 ft plus 6 ft plus 20 ft plus 6 ft plus 14 ft), and 58 LF for the double partition (8 ft plus 7 ft plus 14 ft, all doubled), totaling (60 LF plus 58 LF) 118 LF. Using a single sill plate and a double head plate is 3 times 118 LF. The studs are on 16-in. centers:

$$118\text{--}0 \text{ at } 16 \text{ in. c-c} = 89 \text{ studs}$$
$$+ \text{ 11 walls, 1 for each end} = 11$$
$$+ \text{ Closet studs, 2 at 6 each} = \underline{12}$$
$$112 \text{ studs}$$

40 LF is added for the sills and plates at the two closets. The second floor is similar to the first. Following the above calculations, the low 4-ft partitions allow for 310 LF of bridging (for 306 LF of partition). An extra 16 LF of studding will also be needed for each doorway; 16 LF of additional material is a good average to allow for framing around ordinary door openings.

The blanket insulation is slightly different in height on the two floors, being 8–2 for the first and 8–0 for the second. The 15 percent waste allows for weaving the material around the double studding.

The interior

The finish carpentry and miscellaneous rough carpentry for the building shown in W. D. 7.4 and 7.5 were taken off according to the following specifications: Ceilings are to be furred with 1 in. × 2 in. spaced at 12-in. centers; $\frac{5}{8}$ in. × 1 in. plaster grounds are required for base in all rooms except bathrooms, and around all door and window openings. Continuous nailers (2 × 4) are required for kitchen cabinets and 1 × 2 grounds on 14-in. centers behind kitchen cabinets. Plywood ($\frac{5}{8}$-in.) is to be laid over subfloor. The exterior doors are to be 4-panel pine, 2 ft 8 in. × 7 ft × 2 in. with $1\frac{3}{4}$-in. pine frames, double rabbeted, 2-in. staff bead is required for exterior door and window frames. All doors and windows are to be caulked. Interior doors are to be $1\frac{3}{8}$ in., thick solid-core, flush. The interior door frames are to be $1\frac{5}{8}$-in. pine. Door casings are $\frac{3}{4}$ in. × $3\frac{1}{2}$ in., and base $\frac{3}{4}$ in. × 4 in. in all rooms except baths and closets. The window casings are $\frac{3}{4}$ in. × 3 in., stool 1 in. × 4 in., apron $\frac{3}{4}$ in. × 3 in. The windows are ponderosa pine frames, double-hung wood sashes with spiral balances. The basement stairs are to have 2 × 10 treads, 3 × 4 newels, and 2 in. × 2 in. handrails with 2 in. × 2 in. posts (5 per

144 Chapter Seven

handrail), handrails on both sides, and open risers. The main stairs are to have 2 × 12 carriages; risers and oak treads, framed and glued; oak balusters, newels, and handrails. Kitchen cabinets prefabricated. Closet folding doors to be fabric type, 2 ft 6 in. × 6 ft 6 in., with necessary track. A shelf 12 in. wide, a hanging pole, and 1 in. × 6 in. supports required for each closet. Oak thresholds 1 in. × 3 in. required for exterior doors. The eave fascia is to be 1 in. × 6 in.; corner boards 1 in. × 5 in. and 1 in. × 6 in.; drip mold 1½ in. × 1½ in. over all windows. Combination storm and screen doors of aluminum; aluminum window screens are to be "half" screens.

The take-off (interior—W. D. 7.4, 7.5)

1×2 ceiling furring

Floor area				=	1,990 SF
+ Over stairs	2 × 9–0 × 3–8			=	66
					2,056 SF
	1 × 2 at 12 in. c-c			=	2,056 LF
+	4 × 48 ft			=	192
					2,248
				=	2,250 LF

$\tfrac{5}{8}$-in. × 1-in. plaster grounds

Base	=	524 ft
Doors 42 × 20–0	=	840
Window-stool	=	118
Window casings	=	336
		1,818 LF
	=	1,820 LF

1×2 grounds for kitchen cabinets

2 × 3 × 14–0 = 84 LF

2×4 blocking for kit. cabinet

2 × 2 × 14–0 = 56 LF

$\tfrac{5}{8}$-in. plyscore to floor

As subflr.	=	1,990 SF
+ 10%		200
		2,190 SF
	=	2,200 SF

THE INTERIOR

2 × 10 cut stair stringer 12 ft long   =   6 pcs.

2 × 10 stair treads 4 ft long   =   24 pcs.

3 × 4 newel posts 3 ft long   =   4 pcs.

2-in. × 2-in. stair balusters 2 ft long   =   20 pcs.

2-in. × 2-in. handrail 4 × 12 ft   =   48 LF

Rough hardware   =   L. S.

Temporary doors   =   4 each

Set only

   Ext. wood door frame (2 in.)   =   4 ea.

   Int. wood door frame ($1\frac{5}{8}$ in.)   =   10 ea.

   Ext. wood door 2–8 × 7–0 (2 in.) and hardware   =   4 ea.

   Int. flush ($1\frac{3}{8}$ in.) wood door and hardware   =   10 ea.

   Fabric folding closet door 2–6 × 6–6   =   10 ea.

   Wood window frame (av. 18 SF)   =   22 fr.

   Wood D.H. sashes and hardware   =   22 prs.

   Comb. storm and screen door (alum.)   =   4 ea.

   Window screen (alum. $\frac{1}{2}$-screens)   =   22 ea.

   Eaves fascia 1 in. × 6 in.
                    2 (50 ft + 24 ft)   =   148 LF

1 × 5 corner boards
                    4 × 18 ft   =   72 LF

1 × 6 corner boards   =   72 LF

Window drip mold 1½ × 1½ = <u>118 LF</u>

Staff bead 2 in. (caulked)

        Drs.   4 × 20 ft = 80 ft
        Windows   = <u>336</u>
                                   416 LF

1-in. × *3-in. oak threshold 3 ft long* = <u>4 pcs.</u>

Window trim			*Stool*			*Casing*		
A	4 at 6-0 × 4-0		4/7 ft	=	28 ft	4/16 ft	=	64 ft
B	8 at 4-6 × 4-0		8/5	=	40	8/16	=	128
C	6 at 4-0 × 4-0		6/5	=	30	6/14	=	84
D	2 at 3-0 × 4-0		2/4	=	8	2/14	=	28
E	2 at 5-0 × 4-0		2/6	=	<u>12</u>	2/16	=	<u>32</u>
					118 LF ✓			336 LF ✓

Stool 1 in. × *4 in.* = <u>118 LF</u>

Apron and window casings ¾ in. × 3 in. = <u>454 LF</u>

¾-in. × 3½-in. *door casings* = <u>840 LF</u>

¾ × *4 base*

        Living   2 × 50 ft
        Dining   2 × 30
        Kitchen   2 × 24
        Bedr. 3   2 × 48
        Bedr. 1   2 × 48
        Bedr. 2   2 × 40
        Passage   2 × <u>22</u>
                             262 LF × 2
                     = <u>524 LF</u>

Main stair 3-8 wide (14R) complete including newels and rail = <u>2 sets</u>

Kitchen counter cabinets 12 LF = <u>2 ea.</u>

Closet shelf & pole = <u>10 sets</u>

1 × 6 *shelf bearer (8 LF average)* = <u>10 ea.</u>

Notes on the take-off (interior—W. D. 7.4, 7.5)

The ceiling furring could be measured separately. This is very time consuming and unnecessary. Since the ceiling furring is on 12-in. centers, this equates to 1 LF of furring per SF of the ceiling area. The ceiling area is equal to both the first- and second-floor areas plus the second floor over the two stairs. Then add one length of furring for each side of the partition. There are two partitions that are 48 LF that yield 98 LF. The quantity yield for this simpler method is about the same as a detailed take-off.

The plaster ground items are first listed without quantities and then after the various millwork items have been obtained; the totals are entered in the ground items. The grounds around the doors include both sides of all interior doors except the basement door, which is taken off for one side only, both sides of the cabinet doors, and one side of the exterior doorways.

The basement stair is not a millwork item and is taken off with the rough carpentry for labor and material. The lengths given for the various items are reasonably accurate for a stair about 8 ft high.

The rough hardware and temporary doors were not specified, but will be needed. These items should always be included in the carpentry take-off. The rough hardware provides for nails, screws, sandpaper, glue, and bolts for all the carpentry work in the entire building.

The door and window items are straightforward, and it is not necessary to list them separately by sizes. Reference to the exterior take-off provides the essential information: 22 window openings—400 SF, for an average of 18 SF per opening.

The various trim items for the windows are obtained by totaling the requirements for each window and adding them for the total stools and casings; these totals will then provide the quantities for other window items. The stool quantity is also the exterior drip-mold quantity, while the casing quantity equals that of the staff beads. The combined stool and casing total is used for the plaster grounds.

The door casings are the same as the door grounds and staff bead. All doors sized up to 3–0 can be estimated as two 8–0 and one 4–0 length or 20 LF per door. The wood base is totaled for each room by measuring around the walls and keeping a running total. Remember to use even-foot lengths for material over 8 ft and 1-ft multiples for material under 8 ft.

The main stairs should be taken off as a lump-sum price for material and labor. The price should include all carriages, treads, risers, skirt board, balustrades, newels, and handrails. If a millwork subbid is used that excludes any stair item, then this should be included in the rough carpentry section of the estimate.

The kitchen counters can be prefabricated and shipped in two pieces for installation in each kitchen.

The millwork and casework have not been described in great detail in this estimate since most contractors will solicit lump-sum prices from the specialty millwork suppliers for this item. Some contractors may choose to provide labor to install the millwork, which as mentioned earlier should be included in the rough carpentry take-off.

MISCELLANEOUS ROUGH CARPENTRY

Framing lumber is generally measured in board feet, unless otherwise indicated on the estimate. It should be noted on the estimate when using rough lumber for carpentry work, since rough lumber is more expensive to handle than dressed lumber. Specific dimensions must always be checked to ensure that one is estimating what is being specified. For example, if the roof boarding is to be 1-in. finish thickness, then it is necessary to estimate a $1\frac{1}{4}$ in. thick board. A 1-in. nominal thickness will only yield a $\frac{3}{4}$-in.-thick finish board.

Boarding for floors, roofs, and wall sheathing should have an additional allowance for cutting and waste according to the type of board, width, and pattern layout. Tongue-and-groove (T&G) boarding lays $\frac{3}{4}$ in. less than its nominal width. A 1 in. × 6 in. T&G board would lay only $5\frac{1}{4}$ in. or a net loss of $\frac{3}{4}$ in. in a 6-in. span, which equates to a loss of 14 percent in dimensional area.

Square-edge boarding would lay $\frac{3}{8}$ in. to $\frac{1}{2}$ in. less than its nominal width. The amount of waste increases proportionally with the amount of cutting and fitting required by the job.

Exterior sheathing for a wall with numerous openings will require more cutting waste than subflooring. Cutting waste varies from $7\frac{1}{2}$ to $12\frac{1}{2}$ percent.

Hips and roof rafters can be set up in plan as shown in W. D. 7.7. The rise is set square to the hip in plan and the hip is drawn in a flipped-over position. Allowance must be made for both the overhang and the long bottom cut. The hip length can also be obtained by using hypotenuse rule (the square of the hypotenuse equals the sum of the squares of the two sides).

Laminated wood arches are prefabricated timber arches which are typically shipped to the jobsite for erection. Laminate arches should be taken off by the piece and fully described. Sketches on the estimate take-off sheet are invaluable for these descriptions.

Laminated wood beams, purlins, etc., are similar to the wood arches in that they are typically manufactured off site and shipped to the site for installation. Subbids for these materials should include all necessary bolts, angles, plates, and hardware required for the installation. The estimator should check the subbid to ensure that the hardware has been included. If the hardware is not included, a take-off needs to be made under a separate heading. The estimate should consider the number of pieces, tonnages, sizes, etc., of the laminated wood member and add equipment for off-loading, transportation and erection if not included in the subbid.

Wood roof trusses that are to be fabricated on site should be estimated for material requirements, connectors, fabrication, handling, hoisting, and erection. Erection devices will include items such as cranes, miscellaneous rigging, guying, temporary anchors, and flagger.

Stud framing is usually reduced to board feet. However, if the quantities of material are relatively small and the stock items are also small, it would be wise to list each item separately or at least in linear feet. The linear feet listing could act as a warning to the estimator that the material could be more expensive than standard unit prices for bulk material.

Wall boarding such as plaster board, pressed hardboard, and cementitious board is measured in square feet. If composition board is to be covered by some other finish material, the waste should not be too great and a 10 percent allowance will probably suffice. The finish material waste may be greater. Panels in window walls, for example, might be specified as $1\frac{1}{2}$ in. cementitious board, which is only available in sheets 4 ft wide. Panels slightly more than 2 ft in both dimensions would involve a waste of almost half of each sheet. Cutting 4 ft × 8 ft sheets of cementitious board for panels 2 ft 3 in. wide would mean wasting 1 ft 9 in. out of 4 ft. The result is a considerable quantity of scrap material left over that must be either shipped to a storage yard or to a disposal site. In either case additional costs will be incurred by the contractor to remove the material.

MISCELLANEOUS FINISH CARPENTRY

Moldings are taken off in lineal feet, with each member measured separately. Some estimators take off multiple member pieces as one item. This is not a very satisfactory procedure unless the various pieces and sizes are noted. If there are too many pieces to list individually, it would be prudent to write the item up as "Eaves cornice, 4-member, see Drg...." Trim that is curved or polygonal in shape should be taken off separately from ordinary moldings. It is expensive to fit moldings that do not meet at a true miter.

Paneling is measured in square feet and separated into groups of 4 ft high, 4 ft to 6 ft, and over 6 ft high. The low dado paneling will cost more to install per square foot than similar paneling 6 ft high. Paneling to be built up on the wall is taken off as separate items, such as plywood, cover strips, and moldings. Prefabricated paneling, on the other hand, is measured in square feet of finished panel, with only the base and cap mold measured separately. First-class hardwood paneling should be carefully described to present an item that can be properly evaluated and priced. Paneling of narrow width should be kept separate from the main-run wall paneling. A narrow width may be considered to be up to 15 in. wide as might occur at window reveals, beams, or around isolated columns.

Cabinet and casework is usually taken off in items by size. Long storage cabinets may be measured in lineal feet of a noted width and height. Casework that cannot be preassembled should be kept separate from assembled casework. One should use common sense in determining whether fully assembled items can be installed on the project.

Cutouts in countertops must be taken off as labor items, if not included in the millwork item. Usually cutouts for sinks, grilles, etc., will need to be made on the job with the millwork setup. These cutouts should be considered as an extra cost to set the cabinets.

Window walls are taken off in square feet, fully described. Any trim items that are not part of the window subwork must be taken off separately. Quite often it is best to take off separate items for window walls; trim and cover molds, and sashes for the setting estimates. There are many types of wood window walls; some are prefabricated units, others are simple job-constructed posts and frames. The take-off should employ the best method according to the details shown on the drawings.

Glazed wood partitions and screens are measured in square feet. Preassembled partitions and job-assembled partitions are taken off as separate items. All stud framing, blocking, and backing material for job-assembled partitions should be taken off as rough carpentry.

Chapter 8

Alteration Work

Alteration work should always be taken off separately from new work. There are two reasons for this: (1) The alteration work could be more expensive than new work of the same type. Face brick for filling in several isolated openings could cost much more per brick to lay than face brick for an entire new wing. Patching the wood base in existing rooms could cost much more for labor in proportion to the same amount of material used to than install new wood base. (2) The alteration work must be examined when you visit the site; therefore, the items involved should be clearly listed for identification and evaluation during the inspection. The most effective method to achieve the inspection is an orderly progression through the building room by room. Identify each item by noting its location on the left-hand side of the take-off sheet and then transfer that notation to the estimate sheet.

Architects cannot always show the alteration work in full detail on the drawings; consequently, a dotted line on the drawings, noted simply "remove partition," could be a 12-in. brick wall plastered on both sides and also a load-bearing wall for the two floors above. Similar conditions could present shoring problems and additional equipment that could not possibly be determined from the drawings. Sometimes the drawings show only the revised layout and details, with a simple note to "remove present walls, floors, etc., as necessary." Further, some alterations might be too complex for all the work involved to be shown on the drawings. If the drawings do not show all the removal work, it may be necessary to visit the site and take off some of the items on site, measuring and describing the various items as you see them.

When going through a building looking at removal and alteration work, there is nothing to be gained by rushing through the rooms. This is one time that speed is definitely not the most important consideration. There are usually many factors to consider that are not shown on the drawings but can be seen on the job by an observant estimator. The finish materials to be patched and matched may be materials not easily obtained. Wood moldings, for instance, might need to be custom-made to match the existing materials.

Removal work can be considerably more difficult than the drawings show or could be seen by a casual observer. In addition to the actual work involved in demolition, the rubbish must be removed and, equally important, working conditions may present problems. Cutting and patching in buildings that are occupied will be very expensive because of dust, interruptions, and possibly night or overtime work.

Every alteration job will present its own problems, which must be given some thought either before taking off the work or before pricing the estimate. The take-off should list and describe the work in such a way that the estimate can be properly priced. If you must prepare the take-off without having seen the site, then list doubtful items with question marks. It is better to have an item listed, even if it is later not required, than it is to miss an item that might cost a considerable amount of money. As an example, if there was nothing on the drawing to show load-bearing walls, then a shoring item should be included for the removal of any basement walls. The item might simply be: "Shore first floor for removal of 8-in. wall below—24 LF." The item could then be evaluated fully when the building is visited.

Alteration work will often necessitate using many lump-sum items in the take-off, either for work that cannot be taken off in detail or for removal work that must be priced by on-the-job evaluation of the labor and equipment required for the operation. Overtime work may need to be priced as a lump sum, consisting of the estimated number of Saturdays and Sundays to be worked for certain phases of the alterations. Trucking of rubbish is another item that cannot always be measured. Examples of items that might be priced as a lump sum while actually looking at them are entrance steps or windows to be removed and sash cords to be replaced "as necessary." Any information available concerning an item should be included in the take-off description. If a retaining wall that is to be removed is clearly shown on the drawings so that it can be measured, then it should be measured and the quantity indicated in the item: "Remove concrete retaining wall—182 CF." If the windows are all shown on the drawing, then an item for a requirement to "replace sash cords as required" should indicate the number of sashes: "Replace sash cords as necessary—82 prs. sashes."

In the renovation of any existing building, there is a growing concern of the many environmentally unsafe products that may exist. One of the more publicized materials is asbestos products that were used for many years as insulation. Removal of asbestos and other environmentally unsafe materials can be very expensive and very time consuming.

UNDERPINNING

Working Drawing 8.1 shows an extension to a present building. The floor of the new basement is to be at grade 82–9; the present basement floor is at grade 86–7. The end wall of the present building is to be underpinned as shown, in lengths of 3 ft with 6 ft between the openings. Excavation, except as required for underpinning, may be considered as having been previously taken off. The bulk excavation, however, would not have been taken off all the way up to the old wall because it would be unsafe to do so. The machine excavation should include the entire space up to the old wall down to approximately 2 ft above the present floor and step out to approximately 3 ft from the old wall. The remainder from grade 88–7 at a width of 3 ft for the entire 36-ft length would require careful excavation in sections allowing the underpinning to proceed. The specific method of excavation would depend on job conditions, soil conditions, moisture, and the condition of the old wall.

The take-off (W. D. 8.1)

UNDERPIN END WALL

Excavate adjacent to end wall (part machine)

$$36\text{-}0 \times 3\text{-}0 \times 6\text{-}10 = \underline{\underline{738}} \text{ CF}$$
$$= \underline{\underline{28}} \text{ CY}$$

Hand exc. for underpinning (in 3-ft lengths)

$$36\text{-}0 \times 1\text{-}2 \times 3\text{-}6 = \underline{\underline{147}} \text{ CF}$$
$$= \underline{\underline{5\tfrac{1}{2}}} \text{ CY}$$

Conc. 3,000 psi (Hi-Early) in underpinning (3-ft sections)

$$36\text{-}0 \times 1\text{-}2 \times 3\text{-}6 = 147 \text{ CF}$$
$$36\text{-}0 \times 0\text{-}6 \times 0\text{-}6 = 9$$
$$\underline{\underline{156}} \text{ CF}$$
$$= \underline{\underline{6}} \text{ CY}$$

FORMS (one face only)
$$\underline{\underline{126}} \text{ SF}$$

End stops
$$1\text{-}2 \times 3\text{-}6 - \underline{\underline{13}} \text{ pcs.}$$

$\tfrac{5}{8}$-in. × 12-in. dowels to foundation walls = $\underline{\underline{12}}$ pcs.

Cut off footing projection 6 in. × 12 in. = $\underline{\underline{36}}$ LF

Notes on the take-off (W. D. 8.1)

The excavation adjacent to the wall is noted "part machine" requiring some hand work. The item should be evaluated after one has seen the site and priced accordingly. Let us assume that the excavation would be 70 percent machine work and 30 percent hand work and that the machine work is worth $1.40 per CY and the hand work $6.25 per CY. The item would then be priced as follows:

$$\text{Machine} \quad 70\% \times \$1.40 = \$0.98$$

$$\text{Hand} \quad 30\% \times \$6.25 = \underline{1.87}$$

$$= \$2.85 \text{ per CY}$$

If machine excavation is given to an excavation contractor, then the pricing would be entered as $1.87 per CY on the labor side of the estimate and $0.98 per CY on the material side. If the machine excavation is self-performed by the general contractor's crew, the machine item should be broken down into labor, equipment, and fuel. This procedure would split the machine item into $0.24 for labor and $0.74 for material. The entire excavation item would be priced at $2.11 per CY for labor and $0.74 per CY for material for a total of $2.85 per cubic yard.

Hand excavation for the underpinning is taken off at 1–2, not at the neat 1–0 that is shown on the drawing. It would be difficult and expensive to excavate under an old wall to an exact line. Therefore, it is necessary to allow additional width for both excavation and concrete to obtain realistic quantities.

The formwork is noted as "one face only" to serve notice that this will be expensive formwork to hold in place. The end stops are to contain the concrete at each 3-ft length of formwork.

The concrete item includes the small 6 in. × 6 in. offset. Cutoffs of footing projection and reinforcing in the old wall are as shown on the drawing.

EXCAVATION AND CONCRETE FOR PIPING WORK

Working Drawing 8.2 shows new sanitary drainpipe for a existing building. The piping may be considered as part of the plumbing work. The general contractor's work will consist of breaking out the concrete floor, excavating, backfilling, replacing the floor, and removing the rubbish.

The take-off (W. D. 8.2)

$$\begin{array}{llllll}\text{Machine} & 70\% & \times & \$1.40 & = & \$0.98 \\ \text{Hand} & 30\% & \times & \$6.25 & = & \underline{1.87} \\ & & & & = & \$2.85 \text{ per CY}\end{array}$$

EXCAVATION, ETC., IN PRESENT BAST. FOR MECH. TRADES

Hole through foundation wall for 4-in. soil pipe & make good = <u>1 ea.</u>

Cut conc. floor (4 in.) before breaking out

$$2\,(28\text{--}0 \;+\; 43\text{--}0 \;+\; 7\text{--}0 \;+\; 6\text{--}0) \;=\; \underline{168}\text{ LF}$$

Break out 4-in. conc. floor

$$84\text{--}0 \times 3\text{--}0 = \underline{252 \text{ SF}}$$

Exc. gravel bed (& stockpile)

$$84\text{--}0 \times 3\text{--}0 \times 0\text{--}6 = \underline{5 \text{ CY}}$$

Exc. for plumber

$$84\text{--}0 \times 3\text{--}0 \times 2\text{--}8 = \overline{\underline{672 \text{ CF}}}$$
$$= \underline{25 \text{ CY}}$$

Backfill (compacted) = $\underline{25 \text{ CY}}$

Replace gravel bed = $\underline{5 \text{ CY}}$

Concrete patching floor

$$252 \text{ SF} \times 0\text{--}4 = \underline{3 \text{ CY}}$$

Float & trowel floor = $\underline{252 \text{ SF}}$

Remove rubbish = $\underline{\text{L. S.}}$

Clean up basement = $\underline{\text{L. S.}}$

Notes on the take-off (W. D. 8.2)

The items are simple enough; the above example is probably most useful in showing how to list the requirements in an orderly manner. The hole through the foundation wall includes the cost for finish patching. The cutting and excavation items allow 1 ft extra on the 42-ft dimension for working space. This 3-ft-wide run will also provide the working space necessary for the other three dimensions. The excavation of the gravel is taken off separately, because that material will be stored for the backfilling requirements.

Estimating items will be needed for trash removal and cleanup.

The estimator for the general contractor should always review all sections of the plans and specifications to determine that all items of work have been covered. As an example, the mechanical and electrical contractors bid their respective work items. Both subcontractors will have equipment that must be installed on concrete pads for protection, vibration, etc. During construction each subcontractor will probably declare that concrete work is the responsibility of the general contractor and not the subcontractor. Disputes occur, time and money are lost. Avoid the headaches, review the documents and confirm that the work is covered and include it in the estimate.

REMOVING WALLS (SHORING)

Working Drawings 8.3 and 8.4 show two building walls to be enlarged. Part of the end wall (elevation 1–1) and front wall (elevation 2–2) is to be removed. New lintels, beams, and columns are to be installed to support the floor and roof load and connect the structural steel framing of the addition. A new doorway is to be installed at the basement level. The basement windows are to be closed with common bricks (12 in. thick). The four semicircular windows on the main floor are (4–4 × 9–6) with wooden frames and sashes.

The structural-steel contractor is to supply and erect all structural-steel columns and beams shown on elevation 1–1. The general contractor is to erect lintels A, B, C, D, E, and set the column base plates, beam bearing plates, and anchor bolts. Nonshrink grout is to be used over the lintels. The two fireproofed beams shown on column line F9 are from the new wing and are not part of the alteration work.

Before alteration begins, a dustproof partition is to be installed in the existing building, extending from the floor elevation 286–0 to the ceiling, a height of 22 ft, running from line H to line F on the drawing. A similar partition 11 ft high is to extend from line F to line 11. In addition, the building must be protected from the weather by protecting the outside face as necessary during construction.

The take-off (W. D. 8.3, 8.4)

ALTERATION WORK

2 × 4 *frmg. temp. partition*

End	3	×	26 ft	=	78 ft
	2	×	24	=	48
	21	×	12	=	252
	21	×	10	=	210
Front	2	×	24	=	48
	20	×	12	=	240
					876 LF
				=	585 BF

REMOVING WALLS (SHORING)

⅜-in. plywood to temp. ptn. (taped joints)

$$
\begin{array}{rcrcl}
25\text{–}8 & \times & 22\text{–}0 & = & 565 \text{ SF} \\
23\text{–}0 & \times & 11\text{–}0 & = & 253 \\
& & & & \overline{818} \\
& + & \text{Waste } 10\% & & 82 \\
& & & & \overline{900 \text{ SF}}
\end{array}
$$

Scaffolding for removing walls

$$
\begin{array}{rcrcl}
26\text{–}0 & \times & 30\text{–}0 & = & 780 \text{ SF} \\
26\text{–}0 & \times & 22\text{–}0 & = & 572 \\
& & & & \overline{1{,}352 \text{ SF}}
\end{array}
$$

Remove bast. window 4–4 × 2–9 = <u>3 ea.</u>

Remove window sashes (approx. 4–0 × 3–6) *& store for reuse* = <u>4 prs.</u>

Remove wood window frame 4–4 × 9–6 *& store for reuse* = <u>4 fr.</u>

Cut pocket for column seat 18 in. × 12 in. × 12 in. = <u>2 ea.</u>

Cut pocket for column seat 18 in. × 18 in. × 12 in. = <u>1 ea.</u>

Cut pocket for column seat 48 in. × 12 in. × 48 in. = <u>1 ea.</u>

Conc. col. seats   *FORMS*

$$
\begin{array}{rrcrcrcl}
2/ & 1\text{–}6 & \times & 1\text{–}0 & \times & 1\text{–}0 & = & 3 \text{ CF} \\
& 1\text{–}6 & \times & 1\text{–}6 & \times & 1\text{–}0 & = & 3 \\
& 4\text{–}0 & \times & 1\text{–}0 & \times & 4\text{–}0 & = & 16 \\
& & & & & & & \overline{22 \text{ CF}} \\
& & & & & & = & 1 \text{ CY}
\end{array}
\quad
\begin{array}{r}
10 \text{ SF} \\
6 \\
40 \\
\overline{56 \text{ SF}}
\end{array}
$$

Set col. base plate & U-bolts = <u>4 ea.</u>

Shore 12-in. wall for cutting new doorway 4–0 × 8–0 = <u>L. S.</u>

Remove foundation wall for new doorway = <u>32 CF</u>

Set 10-in. channel lintel 5–4 long = <u>2 pcs.</u>

Patch foundation wall around 4–0 × 7–6 *opg.* = <u>1 opg.</u>

Cut beam pocket in brick wall & set 8 × 12 *base plate & make good* = <u>2 ea.</u>

Cut chase 4 in. deep for 12-in. channel lintel
$$2 \times 26 \text{ ft} = \underline{52 \text{ LF}}$$

Cut chase for 10-in. channel lintel
$$2 \times 25 \text{ ft} = \underline{50 \text{ LF}}$$

Cut holes through 4-in. brick for $1\frac{1}{4}$-in. pipe spreaders = $\underline{20 \text{ ea.}}$

Set inner 12-in. channel lintel = $\underline{26 \text{ LF}}$

Set outer 12-in. channel lintel = $\underline{26 \text{ LF}}$

Set inner 10-in. channel lintels (3 pcs.) = $\underline{26 \text{ LF}}$

Set outer 10-in. channel lintels (3 pcs.) = $\underline{26 \text{ LF}}$

Nonshrink cement dry pack over lintels = $\underline{104 \text{ LF}}$

Remove 12-in. ext. brick wall
$$\begin{aligned}
25\text{--}8 \times 15\text{--}9 &= 404 \text{ SF} \\
9\text{--}9 \times 1\text{--}8 &= 16 \\
9\text{--}6 \times 7\text{--}8 &= 73 \\
15\text{--}0 \times 7\text{--}4 &= \underline{110} \\
&= \underline{\underline{603 \text{ SF}}} \\
&= \underline{603 \text{ CF}}
\end{aligned}$$

Cut chase 4 in. $\times$ 6 in. in top of brick wall for slab seat (D-D) = $\underline{21 \text{ LF}}$

Face-brick patching
$$\begin{aligned}
3\text{--}0 \times 7\text{--}0 &= 21 \text{ SF} \\
&= \underline{150 \text{ pcs.}}
\end{aligned}$$

Common-brick patching
$$\begin{aligned}
21 \text{ SF} \times 0\text{--}8 &= 14 \text{ CF} \\
3/4\text{--}4 \times 2\text{--}9 \times 1\text{--}0 &= \underline{35} \\
\underline{\underline{49 \text{ CF}}} \times 20 &= 980 \text{ pcs.} \\
&= \underline{1 \text{ M}}
\end{aligned}$$

Tarpaulins & protection for bldg. while open (approx.) = $\underline{700 \text{ SF}}$

Truck away rubbish (approx.) = $\underline{10 \text{ loads}}$

Attend on struct. steel sub. erecting 4 cols. = $\underline{\text{L. S.}}$

Generally make good where disturbed = $\underline{\text{L. S.}}$

Notes on the take-off (W. D. 8.3, 8.4)

The most important consideration in this take-off is that the work be planned to avoid costly shoring of the wall load bearing. To avoid shoring, the lintels for the chase should be installed separately before removing the wall below.

The temporary partition is taken off as 2 × 4 studs on 16-in. centers. The bottom studs are 12–0 with 10–0 over for a total height of 22 ft. The three pieces of 2 × 4 shown at 26 ft are for the stud plates. Forty-eight linear feet of studs are added for intermediate bridging and plates. The total length of the two partitions (9–6 plus 16–0) less 2–6 for a work access setback.

The plywood note, "taped joints," is for the specified dustproof partition. Consideration could be given to salvage the temporary partition materials.

Sufficient scaffolding for an entire side, due to the work required, is planned from the basement floor to a height of 2–0 above the lintels.

The reusable windows must be handled with care to prevent damage and replacement.

Estimate items are included for removing beam plates, column bases, and concrete pads per the drawings. The new basement doorway is itemized, although an experienced estimator might handle it with one lump-sum price. For example, "Shore, set lintel, cut door opg. 4–0 × 7–6 and finish."

The cost of setting lintels B, C, D, and E includes several items: cut chases, set inner lintels, set outer lintels, and dry pack. The chase at E–9 is estimated for a 25–0 length less 12 in. for the corner. Elevation 1–1 requires 9–6 and elevation 2–2 requires 15–0.

The face-brick item is for patching the window opening at line "F". The common brick is for patching basement windows.

Tarpaulins are included to protect the open area until the new wing is finished.

There are always numerous items that cannot be anticipated in alteration and rehabilitation projects. Therefore, it is wise to include items to cover the unforseen conditions that may occur. A few examples are unanticipated cutting and patching for the structural steel erection, coordination of demolition work and new work, patching, and finishing.

REMODELING

Working Drawing 8.5 shows alteration work in a building that is to be remodeled to provide a canteen and two new offices. The following information was taken from the specifications and supplements the drawing: the floor to underside of slab height is 9–9, the finish hardware allowance is $280.00, plastering is not included in the contract, linoleum is not included in the contract, hollow metal door frames are to have transoms, single doors are 3–0 × 8–9, double doors are 5–4 × 8–9, the new entrance doors are to have 2 in. × 2 in. staff bead surrounds on the exterior and 1 in. × 6 in. interior casings and the ceiling height is 8–9 in all rooms.

Chapter Eight

The demolition work has been itemized room by room to allow easy identification and a minimum of time loss during the site visit.

The take-off (W. D. 8.5)

ALTERATIONS

Ext.   *Shore 12-in. wall for cutting new opg.* 4–0 × 8–9 = <u>1 opg.</u>

Remove 12-in. masonry to form door opg. 4–0 × 8–9 & set 3 angle-iron lintels = <u>1 opg.</u>

Waterstruck-face-brick patching
23–0 × 0–8 = 16 SF
= <u>120 pcs.</u>

Common-brick patching = <u>240 pcs.</u>

Set. ext. door frame 4–0 × 8–9 = <u>1 fr.</u>

Set ext. wood door & hardware = <u>1 pr.</u>

2 × 2 *staff bead* = <u>26 LF</u>

1 × 6 *door casing* = <u>26 LF</u>

Remove 1 pr. wood doors + H. M. fr. = <u>L. S.</u>

Cut new doorways 3–0 × 8–9 *in 8-in. brick ptn. & set lintel*
= <u>2 opgs.</u>

2 × 4 *stud ptn.*
3 × 16–0 = 48 ft
13 × 10–0 = 130
<u>178 LF</u> = <u>180 LF</u>

Remove 1 pr. doors & frame = <u>L. S.</u>

Remove single door & frame = <u>1 dr.</u>

Cut doorway 5–4 × 8–9 *in 8-in. wall & set lintel* = <u>1 opg.</u>

Cut doorway 4–0 × 8–9 *in 8-in. wall & set lintel* = 1 opg.

Remove 6-in. terra cotta tile ptn.

$$22\text{-}4 \times 9\text{-}8 = \underline{216 \text{ SF}}$$

Remove wood base = $\underline{144 \text{ LF}}$

Remove wall plaster for new dado tile

$$88\text{-}0 \times 4\text{-}0 = \underline{352 \text{ SF}}$$

5-in. × *8-in. glazed facing tile dado, 2 in.*

	88-0 × 3- 7	=	316 SF
Less	14-0 × 0-10	=	12
			304 SF
		×	3.35
			1,018 pcs.
	+ 5%		52
		=	1.070 pcs.

Extra for Group 1

Drs.	4 × 6	=	24 pcs.
Window	2 × 1	=	2
			26 pcs.

Extra for Group 2

Sill	14 pcs.
Cap	70
Base	84
	168 pcs.

Extra for Group 4

Base miters	4 pcs.
Sill miters	2
Jambs — dr. 4 × 2	8
Jambs — window	2
	16 pcs.

Clean & point fcg. tile = $\underline{304 \text{ SF}}$

Remove window stool & apron = $\underline{14 \text{ LF}}$

Cut & fit window jamb casing to new tile dado = $\underline{\text{L. S.}}$

Patch floor as necessary

$$26\text{-}6 \times 22\text{-}4 = \underline{592 \text{ SF}}$$

Reset H. M. door frame 4–0 × *8–9* = <u>1 fr.</u>

Reset H. M. door frame 5–7 × *8–9* = <u>1 fr.</u>

Set new H. M. door frame 4–0 × *8–9* = <u>2 fr.</u>

Patch & reset wood int. door & hdwr. = <u>5 drs.</u>

Common-brick patching

(8 in.)	2/	6–0	×	9–8			
(8 in.)		4–8	×	9–8	22–0 × 9–8 =	213 SF	
(8 in.)	8/	0–8	×	9–8	×	14	
						2,992 pcs.	
					=	3 M	

Sand wood floors

2 & 3	2/ 14–6	×	11–9	=	341 SF
5	35–6	×	13–2	=	469
5	26–6	×	22–4	=	592
					<u>1,402 SF</u>

Finish hardware allowance = <u>$280</u>

General cut & patch & make good = <u>L. S.</u>

Remove rubbish = <u>L. S.</u>

Trucking = <u>L. S.</u>

Notes on the take-off (W. D. 8.5)

The new entrance has been itemized. It is estimated that 8 in. of patching will be required around the door. The common brick is 8 in. thick. Consequently, the quantity should be twice that of the face brick. The interior demolition and cutting of the new doorways have been itemized. The masonry patching and door and frame setting will be included in the overall estimates, respectively.

In the new canteen, the plaster removal item is taken off at 4 ft in height, allowing working space for the new facing tile dado. The 88 LF perimeter is twice the sum of 26–6 and 22–4, less the two doorways. Two courses of facing tile are deducted at the window (14 ft × 10 in.). The 5 × 8 tile courses $5\frac{3}{8}$ in. × 8 in. or 3.35 tiles per SF. The locations of the specials in each group are noted on the take-off sheet. The Group 1 items are for the vertical jambs excluding the base and cap pieces, which are in Group 4.

The common-brick patching allows 4 in. at each jamb of the old doorways, 5–4 plus 2 times 0–4 gives 6–0 and 4–0 plus 2 times 0–4 in. gives 4–8. Cutouts have an-8 in. allowance at each jamb. Patching will be needed at all cutout jambs.

The last three items, general cut and patch, remove rubbish, and trucking can be expensive, an allowance should be made.

MISCELLANEOUS ITEMS

Demolition

Major demolition work, such as tearing down or removing a large part of a building, is usually handled by a wrecking company. Wrecking contractors will usually include all the demolition work down to ground level but exclude the removal of foundation walls and floors below ground level. The estimator must determine the extent of demolition included in the subbid. If the work below ground is not included, it must be taken off by the general contractor. Foundation walls should be measured either in square feet or cubic feet and ground slabs in square feet. If an existing basement is of appreciable size in proportion to that of the new building, the excavation should be adjusted to allow for the existing hole. If an existing basement is to be filled in, an item for fill should be included.

Shoring

If walls or columns that support floors or other loads are to be removed, it is usually necessary to add an item for shoring. For example, if the 43-LF end wall of a 3-story building is to be removed, with the floors and roof to remain, then the take-off should include the following item: "Shore floors and roof through 3 floors for removal of end wall—43 LF."

Partitions and interior walls

If removal of partitions is to be part of the alteration work, an item for patching and finishing the floor will probably be required. The item would be measured in lineal feet. For example, "Patch and finish floor after removal of ptns—…LF."

Roofs adjoining new roofs

New and existing drawings should be examined to determine the roofing work. There will probably be removal of the eave cornices and patching of the old roof at the junction with the new and roofing may need to be removed and replaced. All requirements should be itemized with descriptions.

Occupied buildings

In taking off alteration work for a building that is to be occupied during remodeling, the working conditions must be carefully studied. The limited areas available to the contractor may necessitate additional handling of materials. Many factors must be considered in estimating alteration work: confined working space, protection for the public, lack of space to store material, dust screens, temporary ramps, etc. Any item or condition that could cause an expense should be included in the take-off and fully described.

Chapter 9

Job Overhead

The job overhead or general conditions sheet of the estimate is relatively simple to write up, but can be very difficult to price. It requires a clear understanding of the distinction between job overhead and home office overhead. The cost of maintaining the contractor's home office is normally not included in the job estimate. The job overhead items that are carried in the bid are direct costs to that job. It could include such items as trucking to and from a company storage yard for the job, but not items such as rent and taxes on the yard. Vehicles used exclusively on the job could be charged as direct job overhead, but staff cars used for visiting the job would be more properly charged to home office overhead. The estimator must have a clear understanding of the company's overhead policy to assure accurate coverage in the estimate.

After determining the items for direct job costs, decisions will be required by the company to include supervisory personnel as an overhead or as a cost of work items. Advantages and disadvantages exist in both.

An estimator, after having priced many jobs, will develop an idea of the overhead requirements for a particular project based upon project similarities, historical data of past performances, and profitability. Smaller projects typically require a greater percentage of overhead than larger ones. These preconceived percentages should only be used a guide. One must in every estimate develop an overhead sheet based upon the work to be performed, the location, the duration, and the degree of difficulty. Then and only then can one make the best judgment.

BONDS AND INSURANCES

The cost of the payment and performance bonds is determined by the current rate or formula established between the contractor and the surety company. A typical rate schedule may resemble the following:

On the first 100,000 the rate is 10.00/1000.00

On the next 2,500,000 the rate is 6.50/1000.00

On the next 2,500,000 the rate is 5.25/1000.00

On the next 2,500,000 the rate is 5.00/1000.00

On everything over 7,500,000 the rate is 4.70/1000.00

The rates in this example are for demonstration purposes only; the estimator must rely upon the company's surety to provide the proper rate schedule.

The cost of the bond for an estimate cannot be accurately calculated until the estimate is finalized, since the bond cost is based upon the total project cost. Most contractors establish a summary sheet for all the costs and minor adjustments are made here.

Bid bonds

Bid bond cost may be carried as a home office overhead item and not a job overhead item. When a client requires that a bid bond be furnished with the bid, the estimator should confirm with the surety company that a bid bond can be supplied, since this document will be a requirement of the bid proposal package.

Permit bonds

Some local authorities require a contractor to file a permit bond to indemnify the authority against any costs arising from the contractor using public streets, crossing curbs, connecting to public utility lines, etc. Permit bond costs should be confirmed with the surety company. A single permit bond usually covers a contractor for work within a particular town or city during that year.

Insurances

There are several types of insurance that may be the responsibility of the contractor: worker's compensation, public liability and property damage, contingent liability, and hold harmless. The last two types have standard rates depending on the amount of coverage and can be priced on the overhead sheet without difficulty. The rates for worker's compensation and for public liability and property damage vary according to the trades involved and the state in which the work is performed. There are hundreds of different classifications, each with its own particular rate, but we are concerned with only a few of them. Because of the wide variation in rates, many contractors either carry these insurance items on the applicable trade sheets or use the labor totals from those trade sheets to compute a total insurance item for inclusion on the overhead sheet. Other contractors depend on their accounting departments to give the estimators insurance costs as percentage figures of payroll costs for particular types of jobs.

Builder's risk insurance

Builder's risk insurance premiums are based on either the amount of the contract or the value of the structures excluding excavation work. All insurance requirements should be read carefully. In many contracts, the owner may elect to provide portions of the insurance requirements.

Health and welfare funds

Payments to the various trade health and welfare funds can add 5 percent or more of the payroll. This can be handled in any one of three ways: (1) The health and welfare contribution rate can be added to the hourly wage rate as part of the gross figure used to establish the unit prices. This method may be unsatisfactory, since health and welfare rates vary from district to district and in many areas there are none. (2) The health and welfare item can be entered at the end of each trade sheet in the estimate, with the total labor cost expressed in hours and the health and welfare cost computed and added into the sheet. This is a satisfactory way to handle the item. (3) Health and welfare contributions can be expressed as a percentage of the payroll total and computed as a single item or as a series of items trade by trade and carried either on the overhead sheet or on the trade sheets. This method is quite satisfactory if the payroll percentage is calculated accurately and meets the labor rates and the health and welfare rates for the area.

Pension funds

Contributions to the pension funds of labor unions are handled in the same way as those to the health and welfare funds. Should you be bidding a job in an area where the fringe benefits do not currently apply, you may want to check the expected trend since benefits may well become a cost factor before the completion of your job.

Government payroll taxes

This item covers the employer's contributions for social security (F.I.C.A.), federal unemployment tax, and state unemployment tax. The three items can be combined and expressed as a percentage of the payroll, and the estimated amount computed. State unemployment tax rates vary from state to state. The company accountant should provide the estimator with the appropriate percentages to add for all taxes on labor as well as materials.

FIELD OVERHEAD

Superintendent

The superintendent should always be carried on the overhead sheet. Whether the item should be carried at full or partial salary for the project duration depends on the estimator's analysis of the needs of the particular job.

Supervisors

The supervisor's salary(s) can be carried as an overhead item or as a labor item in the performance of the work. The latter is more desirable in that it does provide historical data that can be used by the estimators in future work. It requires that accurate records be kept to ensure that each supervisor is properly charged to specific work items. If the field staff does not keep good records then all supervisory personnel should be included as overhead.

Engineering

For large projects, it may be necessary to carry an engineering staff for site layout and grades for the entire period. Other jobs might justify only a few weeks for an engineer to lay out the building and the foundation work. If the contract calls for a registered engineer to lay out the building, the cost of the layout should be carried on the overhead sheet.

Timekeepers and material clerks

For large jobs, it is customary to carry overhead items for the clerical help, stock or toolroom clerk, accountants, etc. Each project should be analyzed for its particular needs.

Watchmen

The estimator and contractor must decide if a full- or part-time security staff will be required. Security costs should be on the overhead sheet.

Job offices and shanties

Temporary portable wooden shacks were the recognized type of building for job offices and toolrooms until recent years. They are still probably the most convenient buildings for the purpose, but are more expensive to dismantle, truck, and reassemble than the office trailers and converted truck trailers that are now seen on construction projects. Trailers can be relocated inexpensively and used as transportation for job site tools. Two secondhand truck trailers could cost less than one large wooden shack and will last many more years.

    The overhead item for temporary buildings will vary according to the type of shack or trailer that you intend to use. It should include all installation, maintenance, transportation, erection, cleaning, heating, and fuel costs.

Trucking

General trucking is an item that often costs much more than the contractor anticipates. Trucking from the yard to the job, back to the yard, to the dump, to pick up odd items, etc., can become a large cost item. It is not easy to determine how much trucking a particular job will require. Some contractors keep a separate record of trucking costs and use the data as a basis for pricing trucking in subsequent estimates. All trucks, pickup trucks, and cars to be used either on the job or for the job should be included on the overhead sheet.

Temporary light and power

If the general contractor is to pay for lighting and power, the item must include the cost of hook up of the temporary service, extension cords, temporary wiring, and monthly utility bills. Check the specifications to verify the utility coverage and costs. There could be a cost sharing with the client.

Temporary water

If the general contractor is to pay for the water used during construction, the local water department should be contacted to obtain the water rates. Very often a flat fee will cover all the water used; in other districts, the water may be metered and quite expensive.

Temporary heat

Temporary heat may be required for drying out a building and keeping it warm enough for satisfactory completion of the finishing trades' work. The heating system of the building is generally used for heating during construction if it is possible, but very often the heating system will not be complete in time to supply heat when the contractor needs it. It will be necessary to provide some alternative heating units. Before pricing the temporary heat, one must decide when the building will need heat. After you have determined the duration for temporary heating and the type of heating to be used, the cost of labor, equipment, and fuel can be estimated.

OTHER OVERHEAD ITEMS

Permits

Building permits are necessary except for some public jobs. The cost of the permit is determined by the local authority. The building department in the town or city where the job is located will give you the permit rates.

Equipment

Some contractors carry items on the overhead sheet for equipment such as hoists, cranes, pumps, and compressors. Any equipment that has not been covered on the detailed estimate sheets should be covered on the overhead sheet and priced for labor and material. Whether you would carry equipment items as overhead depends entirely on how you priced the various trade items. If your unit prices provided for the equipment, then do not price the equipment again on the overhead sheet.

Small tools

Some allowance for small tools should be made on the overhead sheet. Tools such as wheelbarrows, concrete buggies, vibrators, and saws get damaged, wear out, are lost, or mysteriously disappear. The office records of small tools purchased for previous jobs should provide a guide for estimating this item. The number of tools that disappear in the first month of the job will tell you whether you allowed enough for the item!

Clean-up

Cleaning up for the various trades, such as cleaning and stacking formwork lumber or clearing up and removing masonry debris, should be charged against the applicable trade labor items. The labor required for the general and final clean-up must be estimated and carried on the overhead sheet. The item should include the cost of material to be used such as soap, sweeping, brushes, cleaners, and rags.

Glass cleaning

If the contract requires the general contractor to clean glass at the completion of the job, the item should be carried on the overhead sheet. It is often possible to obtain a price for glass cleaning from a company that specializes in this kind of work.

Glass breakage

If the contract requires the general contractor to replace all broken glass, this should be covered on the overhead sheet. Almost every building has some glass breakage during construction and except for vandalism or windstorms, the breakage is usually not covered by insurance.

Winter protection

Although some aspects of winter protection and winter working conditions may have already been considered in establishing the unit prices for the various items of work, some general winter protection may still be required. Covering concrete, heating mortar sand, deicing, snow removal, temporary enclosures, etc., are required in winter work. The amount carried will depend on conditions peculiar to each job that is bid.

Signs

The company sign that a general contractor puts up on all jobs is considered a job expense. If the contract requires a special sign (perhaps one showing the title of the project, with the names of the owner, architect, and engineer), then it should be included in the overhead.

Progress photographs

Many contracts require the general contractor to submit photographs recording the condition of the job at certain stages to the owner or architect. These progress photographs can amount to a considerable expense. Four views of the building, with three prints of each view, to be submitted each month, could cost about $200 per month; for a 16-month project, this could cost $3,200.

Premium time (overtime)

Premium payments are the additional cost of overtime labor. That is, the difference between the base rate and the overtime rate. Most construction jobs will involve some overtime work in addition to that already allowed for in the trade labor items. You may need to pour concrete at the end of the day, so six workers work an extra hour, or they pour through the lunch period. There will always be some unanticipated premium time, so an item should be carried for it. Payroll insurance rates are not paid on premium time, so either carry the item on the material side of the estimate or deduct it from the gross labor costs before computing the insurance items.

Labor wage increases

The probability of wage increases during construction must be considered in bidding a job. If existing labor agreements stipulate increases that will go into effect automatically at a future date, it should be possible to figure your unit prices on the basis of the exact rates. The problem with wage increases occurs when it seems impossible to guess what the trades will ask for, what the contractors will offer, or what kind of a settlement can be made.

If a trade has wage negotiations coming up in the immediate future, it should be possible to determine the local feeling and decide how much of an increase may be expected. That is the type of information that a good estimator will find out when visiting a job site. The estimator should look around the area to see how much work is going on, talk to labor delegates, suppliers, or anyone who might give assistance. The information gained enables an estimator to approach the bid more prepared.

The question of whether there will be sufficient tradespeople available to service the job should also be given some thought. Will it be necessary to import labor at an extra expense? Should you cover the extra expense in the labor increases item?

Travel time

All travel time expenses for trades that have union agreements should be included on the overhead sheet. If the project being bid is in a remote rural area that will require large labor crews to be imported from larger districts, then one must consider the travel expenses involved and include those as well.

Subsistence allowance

If workers have out-of-town expenses, these are usually paid by the contractor. It should not be too difficult to compute this item. Usually only a few key people are sent to out-of-town jobs. Perhaps the superintendent, two or three supervisors, and some workers.

Overhead in general

The more common overhead items have been listed and briefly explained. This listing will not cover all requirements for every possible project. Special jobs will require special considerations and present unusual problems. No arbitrary limit can be established for items required on the overhead sheet. There is no standard overhead sheet. However, there should be an organized overhead checklist established by the estimator. The checklist will not prevent the estimator from missing items that should be included, but it certainly will reduce the number of items that could be overlooked.

Chapter 10

Subcontractors' Bids

Analyzing subcontractor quotations is usually not the straightforward mechanical process that the uninitiated may imagine. The worst enemy in the process is "time." Very rarely will the estimator be able to compare subcontractor quotations prior to the bid day. The most that one could hope for is to receive enough proposal forms from subcontractors indicating their intended coverage of the work. The bid proposals will arrive by mail, by carrier, and by telephone calls. There will be last-minute telephone changes to the bid proposals by the subcontractors. There will be questions that the estimators will have of the subcontractors in reference to the quotations.

After the quotations are received, the estimators must reduce the bids to a common basis to accurately compare one to another and ultimately determine the successful bidder. Note, successful bidder, not necessarily the lowest. If the lowest quotation is considerably lower than the rest of the field, there may be a problem in the subcontractor's estimate. To use a price that is considerably lower is proceeding at risk. Read the proposal carefully, verify compliance with the specifications, determine the proper coverage, and decide if the subcontractor can perform the work. If necessary, contact the subcontractor and verify any questionable item.

The bid day process requires a chief estimator who will handle the summary sheet and will enter the prices. The chief estimator should assign different individuals to handle portions of the estimate and analyze quotations from the respective subcontractors. The team should be prepared to give their analysis to the chief estimator upon request and not in an unorganized manner. This will allow the chief estimator a better opportunity to make quick and rational decisions.

Items such as bid proposal forms, alternates, bonds, insurance, certifications, etc., should be processed as much as practical in advance of bid day. The chief estimator should have an assigned individual prepared to finalize these documents as the information is obtained.

The following items are presented to give an idea of some of typical pitfalls to look for in performing an analysis of the subcontractor bids.

HEATING, VENTILATION, AND AIR CONDITIONING

In reviewing the bids for the heating, ventilation, and air conditioning (HVAC) work, there are many items that can be provided by the general contractor as well as the HVAC subcontractor. The burden is on the general contractor's estimator to verify that these items have been covered in the estimate and not duplicated. Typical items are: (1) excavation for underground lines, tanks, concrete work, etc.; (2) concrete, insulating concrete, concrete hold-down pads; (3) miscellaneous steel pipe hangers and steel supports for HVAC units; (4) intake and exhaust louvers; and (5) temporary heat. It is the design engineer's responsibility to specify the work item and not to determine which contractor performs the work. The general contractor must confirm the coverage and decide who does the work. The above listing is not complete and is used only as an example of items that are quite often overlooked or duplicated in the estimating process.

ELECTRICAL

As with the subcontractor items above, the estimator should confirm estimate coverage of the electrical section of the specifications. Items that can be overlooked are: (1) excavation for underground electrical conduit, (2) backfill and compaction of the electrical excavations, (3) concrete encasements, (4) concrete equipment pads, and (5) temporary electrical for lighting and construction services. These items could be provided by the general contractor or the electrical subcontractor.

STRUCTURAL STEEL

Projects that have both structural steel and miscellaneous steel can create minor problems for almost anyone. A structural steel supplier may consider structural steel as only those items that are truly structural in nature, such as columns, beams, purlins, and structural bracing, leaving roof opening frames, fascia-framing steel lintels, and other items to be supplied by a miscellaneous supplier. Some larger structural-steel fabricators do not fabricate the miscellaneous items. Therefore, automatic coverage does not always occur.

MISCELLANEOUS STEEL AND IRON

The miscellaneous steel on a project can be a very expensive item. Consequently, a very thorough check must be made by the estimator to confirm that all items have been covered. Unfortunately, there is no easy way to do this other than spending several hours making a detailed listing of all miscellaneous steel items that can be found on the project plans and in the specifications. One cannot rely upon the steel suppliers to provide the coverage that is needed. The steel suppliers, both structural and miscellaneous, will make a detailed listing of all items that they can foresee out of necessity. Steel suppliers must make their list in order to make shop drawings for their fabricators. This list will probably be attached to their quotation for two reasons: to make sure that you get what they have quoted and that you get no more. This puts the monkey on the contractor's back to make sure that everything is included. Be warned, the miscellaneous take-off on some projects is very tedious and time consuming. Therefore, do not put this work off until the last minute and try to compare the subbids without the list.

ROOFING AND FLASHINGS

The majority of construction projects require a bonded roof warranty. Because of the bonding requirement, roofing manufacturers require that a roof system be used that has been tested and approved. Likewise the roofing subcontractors, in many instances, must also be approved. Because of these requirements, most of the drawings and specifications for roofing and flashings are good. The estimator should review the roofing requirements and confirm that the subcontractors meet all the requirements.

GLASS AND GLAZING

Items that are sometimes overlooked are glass cleaning and replacement of broken glass. The glass and glazing contractor's position is typically "the glass was clean and unbroken when I installed it last year." Invariably, there will be broken glass and at the time of final completion, the glass will never be clean enough for the client. Consequently, one should have an item in the estimate for breakage and cleaning.

MILLWORK

Subcontractor bids for millwork will vary. Many will bid to fabricate and deliver, while others will bid a complete package to fabricate and install. The bids will need to be compared to determine the best package. A general contractor must decide if he or she has skilled craftspeople available for the work, if there is time for the general contractor's forces to do the work, and if it is economically feasible. Many times the millwork needs to be installed on a piece-by-piece basis and in most cases as late as possible to prevent damage.

MISCELLANEOUS ITEMS

The following are only a few items that are often excluded in subcontractor's quotations: (1) scaffolding, (2) hoisting, (3) clean-up, (4) cutting and patching. There may be advantages or disadvantages to the general contractor for the subcontractor to omit items similar to scaffolding and hoisting. The general contractor may provide these items for all the contractors on the site and in turn reduce the cost of the bid to the client while becoming more competitive. Items similar to clean-up, on the other hand, should never be omitted from a subcontractor's proposal. An estimator can seldom realize the costs involved in keeping a project cleaned up during construction.

SUBBIDS IN GENERAL

The examples mentioned in this section do not represent everything that can cause last-minute problems in deciding who to use or what to use. Each project that is bid will present its own set of circumstances. You may expect to have numerous bidders on each subitem only to find no one has provided a quotation. Typically, this will happen when there is a problem or risk that the subcontractor does not want to deal with. If it is easy, everyone will bid, while no one wants the "hard stuff." These items should not surprise an estimator, since the best way to avoid them is to talk to those subcontractors who have indicated a willingness to bid. If problems exist, these subcontractors will let you know and in turn there may be time to get a clarification from the architect or engineer. Another way that this problem can be minimized is to have your own estimate completed and priced to the best of your ability. It may be possible to add a risk factor to your own estimate and continue with the remainder of the estimate without excessive panic.

If the estimator has made take-offs, listed each item in an orderly manner where it may be rapidly retrieved, made numerous preliminary contacts with some of the subcontractors, and prepared a summary sheet that will cover all the work, then bid day may go smoothly. If the preliminary work has not been done and organized properly, then it is suggested that after "hell day" is over, sit down and think about what happened, list all the things that went wrong, and be prepared the next time. Consistent good bidding is not an accident nor is it the result of a mechanical process. It results from a careful and complete study of each project, an accurate and complete take-off of the work, and the use of all the resources within one's organization as well as the subcontracting industry.

PLUMBING

Check the excavation for utilities (interior and exterior), utilities piping, manholes, and catch basin items in the specifications. Some specifications require the plumber to do all work for storm and sanitary sewers, including the excavation, while others require the general contractor to do the excavation, backfill, manholes, and structures, with the plumber responsible for supplying and laying the pipe. Regardless of specification requirement, it is the responsibility of the general contractor to confirm that all work items are covered. A subbid submitted "as per plans and specifications" can be taken as a definite offer to carry out the work as specified. If a subbid spells out exclusions, those items must be covered elsewhere in the estimate.

The specified requirements for temporary water lines should also be checked for inclusion by the plumbing contractor.

METAL WINDOWS AND CURTAIN WALLS

These two items may be in separate sections of the specifications, but generally the curtain walls, of either metal or one of the combinations of porcelain panels, stainless steel, or aluminum, are fabricated and erected by metal window companies. The specified requirements should be carefully checked against the subbids. Items often included in curtain wall or window sections, but excluded by subbidders, include unloading and protecting material, perimeter caulking, and structural framing. The structural window framing item can be especially troublesome if these items are shown on the architectural drawings but not on the structural drawings. The structural-steel subcontractor may disclaim responsibility for miscellaneous framing not shown on the structural drawings and the metal window subcontractor may contend that he or she is not a steel contractor. The responsibility again belongs to the general contractor.

DOORS AND DOOR FRAMES

Subbids on doors and door frames will normally be for supply and delivery, leaving the installation of these items for either the general contractor or an independent installer.

Chapter 11

Industrial Building

GENERAL CONTRACTOR'S ESTIMATE

Working Drawings 9.1 to 9.4, as augmented by data given hereunder, provide the basic information needed to prepare the general contractor's estimate for a manufacturing building of about 14,000 SF. Because of the limitations set by the drawing sizes, the drawings are much simpler than bid drawings would be. However, between the drawings and the outline specification, there is enough information for estimating, at least as an illustration of how the estimate goes together. The building will be taken off and priced and the final estimate compiled just as if we were doing it in a general contractor's office.

It should be noted that the detailed data are limited to the items normally covered by the general contractor, and no effort has been made to show or describe specialty work or subtrades in any detail. The outline specifications define the extent and limits of the work. For convenience, some descriptions, door schedules, and room finish schedules normally shown on the drawings are included in the text.

Outline specifications

General Conditions

Include performance and payment bond; job office; temporary telephone, light, and power, heat, water; insurances: worker's compensation, PL & PD, automotive; taxes: all payroll taxes.

Do not include building permit; fire insurance; long-distance telephone calls by owner's representative; removal of existing buildings; landscaping beyond grading limitations; exterior utilities 5 ft beyond building; site work not expressly included in specifications.

Excavation

Strip and stockpile 6 in. topsoil to all regraded areas only. Building excavation: bank gravel backfill compacted 95 percent to interior.

Grading cut and fill. Disposal one mile from site.

Grade and shape 40 ft beyond building all around including 6 in. topsoil and lawn seed with lime and fertilizer.

Concrete

All concrete 3,000 psi.

Ground slab 5 in. + 6 in. × 6 in. × 6/6 mesh in flat sheets lapped 6 in.; steel trowel finish; Visqueen vapor barrier lapped 9 in.; $\frac{1}{2}$-in. premolded asphaltic expansion joints at perimeter of all slabs at walls.

Styrofoam insulation 1 in., 2 ft deep to exterior walls of office area and 10 ft 0 in. beyond at both sides, only.

Include exterior concrete platforms—1 @ 10 ft × 6 ft, 4 @ 6 ft × 4 ft—all 6-in. concrete on 6-in. gravel with 6 in. × 12 in. deep curb wall to three sides.

Loading docks: 6 in. concrete on 12 in. gravel with 8 in. × 36 in. concrete foundation wall on 20 in. × 12 in. footing, reinforced similar to main walls and slab.

Reinforcing steel: new deformed bars.

Set bearing plates and anchor bolts for columns.

Utilities

Include 340 LF interior trench excavation average 2 ft 6 in. deep for mechanical trades; compacted backfill.

Masonry

Face brick: allow $350 per thousand for brick stacked at site. Common bond with snapped Flemish header in sixth course. Precast concrete windowsills 7 × $5\frac{1}{4}$ concrete block with filler blocks, concrete brick and specials as required.

Dur-o-wall reinforcing every alternate block course.

Control joints caulked, to exterior walls as noted 'CJ', use "half-block" control joint blocks every other course.

Build in open-web joists and structural steel.

Structural steel

    Open-web joists

Metal deck

  Light and ornamental metal

  Roofing and moisture protection

Caulking

Carpentry

  Roof blocking and nailers pressure-treated.

  Roof framing; 1 in. T&G roof boarding.

  Grounds 1 × 2 unless noted.

  Plaster grounds $\frac{5}{8}$ in. × $\frac{3}{4}$ in.

  1 × 2 furring at 12 in. c-c for wood paneling.

  Temporary doors.

  Plywood paneling $\frac{1}{4}$ in. mahogany.

  Trim clear pine unless noted.

  Bench in toolroom: 30 in. wide, 3 ft high with 1 shelf 24 in. wide under.

  Shipping counter: 26 ft long × 3 ft on 4 in. × 4 in. legs, plus 3 ft × 3 ft wicket gate.

  Shipping counter at wall: 22 ft long × 4 ft × 3 ft

  Shipping tables not in this contract.

  Include also: 2 chalkboards 4 ft × 3 ft in wood frames.

      Closet shelf 15 in. wide with hanging rod (2 closets).

      Sliding wood doors and tracks for closets.

  Kalamein doors

    Hollow metal doors and frames

Wood doors

Overhead doors

Aluminum windows and entrances: Weatherseal DL90 or Frontier aluminum

Glass and glazing

Finish hardware: Allowance for purchase only $4,000

Roll-up grilles, aluminum

Fur, lath, and plaster ⎫
                       ⎬ Including all clg. furring for plaster and drywall
Drywall   ⎭

Quarry tile

Ceramic tile

Resilient flooring

Acoustical work (including clg. suspension systems)

Painting and vinyl wall covering

Carpet: Not in this contract

Building Specialties

Toilet accessories—allowance $1,000 for purchase

Metal office partitions

Metal toilet partitions

Aluminum sliding pass windows

Door louvers

Plumbing: General contractor will provide temporary water, all excavation, and backfill.

Heating, ventilation, and air conditioning: Temporary is by general contractor.

Electrical: Temporary light and power is by general contractor.

Sprinklers: Not in this contract.

Alternate prices

A1 Contractor shall submit an alternate price for the add or deduct to provide 2 in. × 4 in. wood stud partitions 7 ft high with ¼ in. birch plywood both sides, ¾ in. × 3 in. birch base and ¾ in. × 6 in. birch cap, natural lacquer and varnish finish, all in lieu of the metal office partitions in Office No. 2.

A2 Contractor shall submit an alternate price to add two coats liquid hardener to floors in manufacturing, receiving, and shipping areas.

Bids. Bids shall be submitted to Hawker Machine Inc., 40 West St., Maxfield, N.Y., on or before 2 P.M. April 8, 19__. Bidder shall quote on all work as requested, without omissions and with no variation from design or specifications. Bidder shall submit with his bid a bid bond as guarantee of the performance and payment bond.

Bidder shall state the time required for the completion of the contract as counted from the date of contract award, which will be within 20 days of April 8, 19__.

Failure to observe all the bidding requirements may result in disqualification of the defaulting bidder. However, the owner reserves the right to award the contract to any one of the bidders or to reject all bids at his, the owner's, discretion.

FINISH SCHEDULE

Room	Floor	Base	Walls	Clg.	Hgt.	Remarks	Abbreviations
LOBBY	QT	—	Pl/Vin	Ac	9-0	Mat 6-0 × 3-0	Ac $\frac{5}{8}$ in. Acoustic tile 24 × 12
CORRIDORS	VA	Rub	Pl/Pt	Ac	9-0	Face brick E & W	Ac* $\frac{5}{8}$ in. Vinyl-faced acoustic tile 12 × 12
PRESIDENT	Carp	W	Pl/Pt Wood	Ac*	9-0	Pl/Pt N & W Wood S & E	Carp Carpet (NIC)
MANAGER	Carp	W	Pl/Ft Wood	Ac*	9-0	Pl/Pt SE & W Wood N	CB Concrete block
PRES.' TOILET	Cer	Cer	DW	Ac	9-0	2 × 4 Clg. joists	Cer Ceramic tile
TOOLS	Conc	—	CB	—	—		DW $\frac{5}{8}$ in. Gyp. bd./pt.
WOMEN	Cer	Cer	Cer	Ac	8-0	Entry-carp.; Pl/Pt	Pl Plaster
MEN	Cer	Cer	Cer	Ac	8-0		Pt Paint 3c
PLANT TOILET	Conc	—	CB	—	—		QT Quarry tile 6 × 6 × $\frac{1}{2}$
DISPATCH	Conc	—	CB	DW	10-6±	DW ceiling on furring	Rub Rubber base 4 in. coved
SHIP. CLERK	Conc	—	CB	DW	10-6±	DW ceiling on furring	W Wood base 1 × 6
OFFICE #1	VA	Rub	Pl/Vin Wood	Ac	9-0	Pl/Vin SE & W Wood N	Wood Mahogany ply $\frac{1}{4}''$ on 1 × 2 furring at 12 in. c-c
OFFICE #2	VA	Rub	Pl/Vin	Ac	9-0	Metal office ptns. 7-0	VA Vinyl asbestos tile $\frac{1}{8}$ in.
SHIPPING	VA	—	CB	—	—	All interior conc block painted 2 coats	Vin Vinyl wall fabric
RECEIVING	Conc	—	CB	—	—		
MANUFAC.	Conc	—	CB	—	—		

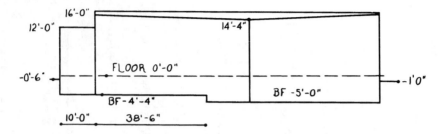

Structural Outline, not to scale

Wall footings 24 in. × 12 in. + 2, No. 4 top & bottom; No. 4 dowels 12 in. long @ 12 in. c-c.

Foundation wall 12 in. + No. 4 @ 12 in. e.w.e.f.

Column footings 4–6 × 4–0 × 1–2, + 5, No. 6 both ways; No. 5 dowels 16 in., 4 per pier.

Column piers 1 ft 2 in. × 1 ft 2 in., + 4, No. 6; No. 3 ties @ 9 in. c-c.

Load-bearing concrete block 12 in. and 8 in.; one course bond beam block with concrete and 2, No. 4 bars, at all three main bearing walls at bar joist bearing. Pilasters 16 in. × 16 in. (16 in. × 4 in. projection) with 6, No. 4 bars, concrete filled.

	Windows		Lintel		
	L	H			
W1, W6	7–4 × 6–8	WF 8 in.	× 6½ in. × 24#	@ 8 ft–10 in. long
W2, W5	15–9 × 6–8	WF 10 in.	× 8 in. × 33#	@ 17 ft– 4 in. long
W3, W4	2–9 × 6–8	L 3@4 in.	× 3½ in. × ⅜ in.	@ 3 ft– 9 in. long

TAKE-OFF INDUSTRIAL BUILDING

Working drawings 9.1 to 9.4

EXCAVATION & SITE

Strip & stockpile topsoil   SF
219–0 × 179–0 = 39,201
131–0 × 10–0 = 1,310
                   40,511 × 6 in. = 20,256 CF
                                    = 750 CY

(continued on p. 188)

DOOR SCHEDULE

Opg. No.	Size	Door	Frame	Lintel	Remarks
B	6–0 × 8–4	Pr. alum-glass	Alum	3 @ 4 × $3\frac{1}{2}$ × $\frac{3}{8}$	Special
C	3–0 × 7–0	SC Flush wood $1\frac{5}{8}''$	HM	PCC 12 × 8	SC: Solid core Glass panel 24 × 18
D	3–0 × 7–0	HM $1\frac{3}{4}''$	HM	PCC 8 × 8	2 hr. rated
E.F.G.	3–0 × 7–0	SC Flush wood $1\frac{5}{8}''$	HM	—	—
H	2–4 × 6–6	HC Flush wood $1\frac{3}{8}''$	HM	PCC 8 × 8	HC: Hollow core Louver 18 × 12
I	2–8 × 7–0	SC Flush wood $1\frac{5}{8}''$	HM	—	—
J.K.L.M.	3–0 × 7–0	HM $1\frac{3}{4}''$	HM	PCC 8 × 8	Monly PCC 12 × 8 Monly louver 20 × 16
N.O.	2–6 × 7–0	SC Flush wood $1\frac{5}{8}''$	HM	—	Louver 18 × 12
P	2–8 × 7–0	HC Flush wood $1\frac{3}{8}''$	HM	—	—
Q	6–0 × 7–0	Pr. HC Flush wood $1\frac{3}{8}''$	HM	3 @ 4 × $3\frac{1}{2}$ × $\frac{3}{8}$	—
R	3–0 × 7–0	HM $1\frac{3}{4}''$	HM	PCC 8 × 8	—
S	3–0 × 7–0	HC Flush wood $1\frac{3}{8}''$	HM	PCC 12 × 8	—

Door Schedule (*Continued*)

T	10–0 × 10–0	Overhead wood	12 in. Chan.	WF 8 × 8 × 31#	12-in. Chan.—Jambs only
U	2–8 × 7–0	HC Flush wood $1\frac{3}{8}''$	HM	PCC 12 × 8	Glass panel 24 × 20
V.W.	2–8 × 7–0	HC Flush wood $1\frac{3}{8}''$	HM	—	—
X	16–0 wide	—	—	—	Full hgt., no door
Y	10–0 × 10–0	Overhead wood	12 in. Chan.	WF 8 × 8 × 31#	12-in. Chan.—Jambs only
AA.	4–6 × 7–0	—	HM	PCC 8 × 8	Cased opening
E1	6–0 × 9–0	Pr. alum. glass	Alum.	3 @ 4 × $3\frac{1}{2}$ × $\frac{3}{8}$	Special 8-in. alum. threshold
E2.E3	3–0 × 7–0	HM $1\frac{3}{4}''$	HM	3 @ 4 × $3\frac{1}{2}$ × $\frac{1}{4}$	2-hr. fire 5-in. alum. threshold
E4.E5. E6.E7.	2–8 × 6–8	HM $1\frac{3}{4}''$	HM	3 @ 4 × $3\frac{1}{2}$ × $\frac{1}{4}$	2-hr. fire 5-in. alum. threshold
E8	10–0 × 10–0	Overhead wood	12-in. Chan.	WF 8 × 8 × 31#	Heavy-duty comm. 12-in. Chan.—Jambs only
E9	16–0 × 10–0	Overhead wood	12-in. Chan.	WF 8 × 8 × 31#	Same as E8

Industrial Building

Industrial Bldg. Collection Sheet

Fndtn. walls 12 in.		6-in. Conc. block					8-in. Conc. block			
						Drs.			Drs.	
3–4	4–0	8–8	14–8	9–4	1–8	S Pr	10–0	14–8	S Pr	
139–0	139–0	10–0	48–0	22–0	18–8	2	26–0	54–0	6	
47–6	59–6	12–0	43–6	22–0		6		54–0		
47–6	59–6	15–0	22–0	23–6		8		108–0	1@6–0×4–0	
49–0	137–0	11–0	6–6	21–6		=			1@4–6×7–0	
		48–0	4–6	89–0						
283–0	395–0		124–6							

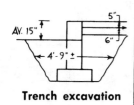

Trench excavation

12-in. Block		4-in. Block		2-in. Block	
9–0	Outs	9–4	Outs	9–4	Out
33–6	1@6–0×8–4	21–6	1@4–0×3–0	9–4	1@4–0×3–0
		21–6		9–4	
	1 Dr 185F	12–0		18–8	
		12–0			
		9–0			
		9–0			
		85–0			

Int. face brick	
9–4	Outs
9–4	
9–4	1@4–0×3–0
18–8	

Wood stud partition

Wall	Studs	Length, ft
6–6 × 10–0	8	8
3–9 × 10–0	4	4
17–0 × 10–0	18	18
3–0 × 10–0	4	4
	34	34

(continued from p. 184)

Grading exc.
SW	=	2–9	39,200 SF × 3–2	=	124,134 CF	
NW		4–0	1,310 SF × 2–6		3,275	
NE		3–6			127,409	
SE		2–3	Minus bldg. 14,630			
	4)12–6		SF × 4 in.		4,877	
	3–1½				122,532	= 4,540 CY

Trench exc. fndtn. walls
Front	283–0 × 4–9 × 3–4	=	4,480 CF	
Rear	395–0 × 4–9 × 3–10		7,192	
			11,672	=	435 CY

Exc. col. ftgs.
5 × 7–6 × 7–0 × 3–4	=	775 CF	
4 × 7–6 × 7–0 × 3–10		805	
		1,580	=	60 CY

Extra cost hand exc. bottom of ftgs.
678–0 × 2–6 × –3	=	424 CF	
9 × 20 SF × –3		45	
		469	= 18 CY

Backfill walls & ftgs.
Total excavated			495 CY	
+ Dock			43	
			538	
− Ftgs., piers	64	⎫		
Walls	68	⎬	137	
Dock	5	⎭		
			401	
+ Compaction 20%			80	
			481	

Interior backfill bank gravel compact 95% = 270 CY
Exterior backfill site material = 211 CY

Hand trim for ground slabs   13,600 SF

Exc. loading docks
Walls	81–4 × 4–9 × 2–9	=	1,063 CF	
Slabs	57–4 × 5–4 × –3		77	
			1,140	= 43 CY

Bank gravel bed, docks
57-4 × 5-4 = 306 SF × 12 in. = 306 CF
+ Compaction 20%   61
                                 ___
                                 367   = __14 CY__

6-in. bank gravel bed, ground slabs
137-0 × 96-0 ⎫
 49-0 × 9-0 ⎭   13,600 SF × 6 in. = 6,800 CF
                                   = 252 CY
+ Compaction 20%   50
                                      ___
                                      __302 CY__

Shape & trim site area
Area stripped   40,511 SF
— Bldg. slab   13,593 ⎫
  Bldg. walls   678 ⎬ 14,631
  Dock   360 ⎭ _____
                           25,880 SF

6 in. topsoil spread & leveled
Site mat'l   25,880 SF = 12,940 CF
+ Compaction 25%   3,235
                             _____
                             16,175   = __600 CY__

Lawn seed + lime + fertilizer   25,880 SF

Disposal surplus exc. mat'l (1 mile)
Grading exc.   4,540 CY
Trench & cols.   495
Docks   43
                           _____
                           5,078
— Backfill (site mat'l)   211
                           _____
                           4,867
+ Bulk up 15%   733
                           _____
                           __5,600 CY__

Allow for pumping (surface water only?)   __L. S.__

Int. trench exc. & backfill, utilities
340-0 × 3-0 × 2-6 = 2,550 CF = __95 CY__

CONCRETE (3,000 PSI THROUGHOUT)

Conc.—wall ftgs.
    678–0 × 2–0 × 1–0   1,356   *FORMS*
    81–4 × 1–8 × 1–0   136   1,356
                                    1,492   164
                                                                         1,520 SF
                          =   56 CY
                                                        2 × 4 Keyway = 760 LF

Conc.—fndtn. walls
    283–0 × 1–0 × 3–4   =   944 CF
    395–0 × 1–0 × 4–0   1,580   2 × 678 ft × 4 ft
                                  2,524   = 5,424 SF
   – Shelf 71–0 × –4 × –8   16
                                  2,508   4-in. Brick shelf 8 in.
                                                                     = 71 LF
                        =   93 CY

Conc.—pier ftgs.
    9 × 4–6 × 4–0 × 1–2   =   7 CY   180 SF

Conc.—piers
    5 × 1–2 × 1–2 × 2–9   =   19   70 SF
    4 × 1–2 × 1–2 × 3–5   19   66
                                      38   136 SF
                        = 1½ CY

Conc.—dock walls
    81–4 × –8 × 2–6   =   5 CY   410 SF

Visqueen vapor barrier
    Ground slab   13,593
    Dock   306
                        13,899
    Laps & waste 10% 1,391
                        15,290 SF

6 in. × 6 in. × 6/6 mesh in flat sheets = 15,290 SF

Conc.—5-in. ground slab
    13,593 SF = 5,664 CF
              = 210 CY

Trowel finish ground slab = <u>13,593 SF</u>

Conc.—6-in. dock slab
    60–0 × 6–0 = 360 SF = 180 CF
                              = <u>7 CY</u>

Float-finish dock slab = <u>360 SF</u>

½ × 6 *Premold exp. jnt.*
    Bldg.   4 × 137 ft = 548 LF
            2 × 49 ft   98
            2 × 105 ft   210
    Dock at bldg.   60
                         ———
                         916
    + Waste   28
                         ———
                         944 = <u>950 LF</u>

6-in. Edge form at dock slab = <u>84 LF</u>

Slab screeds (main floor)
    13,600 SF @ 1 LF/10 SF = <u>1,360 LF</u>

Rub exterior walls
    Allow   449 ft × 1–0 = <u>450 SF</u>

1-in. Styrofoam insulation, fndtn. walls
    Front wing   90–0 × 2–0 = <u>180 SF</u>

Exterior platforms
    Excavation: 11–0 × 7–0 × 1–0 = 77 CF
                   4 × 7–0 × 5–0 × 1–0 = 140
                                        ———
                                        217 = <u>8 CY</u>

    Conc. 6 in. platforms   FORMS
        10–0 × 6–0 = 60
    4 × 6–0 × 4–0   96
                    ———
                    156 SF = 78 CF
        Curb 70–0 × –6 × –6 = 18   70 ft × 1–6 = <u>105 SF</u>
                             ———
                             96 CF
                           = <u>4 CY</u>

Float platforms   = <u>156 SF</u>

Gravel bed 6 in. × 121 SF   = <u>3 CY</u>

Rub curb face   = <u>78 LF</u>

6 × 6 × 6/6 *mesh*   = <u>170 SF</u>

Set 2 anchor bolts and base plate = <u>9 sets</u>

REINFORCING STEEL

								No. 3	*No. 4*	*No. 5*	*No. 6*
Wall ftg.	No. 4	4	×	800 ft					3,200 ft		
Dowels	No. 4	775	×	1 ft					775 ft		
Walls	No. 4	2	×	5	×	800 ft			8,000 ft		
Walls	No. 4	2	×	775	×	3–6			5,425 ft		
Col. ftg.	No. 6	9	×	5	×	4–6					203 ft
Col. ftg.	No. 6	9	×	5	×	4–0					180 ft
Col. ftg.	No. 5	9	×	4	×	1–4				48 ft	
Piers	No. 6	5	×	4	×	2–6					40 ft
Piers	No. 6	4	×	4	×	3–3					52 ft
Piers	No. 3	5	×	4	×	4–6		90 ft			
Piers	No. 3	4	×	5	×	4–6		90 ft			
								180 ft	17,400 ft	48 ft	485 ft

No. 3	180 ft	×	0.376 lb	=	68 lb	
No. 4	17,400 ft	×	0.668 lb		11,624	
No. 5	48 ft	×	1.043 lb		50	
No. 6	485 ft	×	1.502 lb		728	
					12,470	
+	Cut & waste 4%			±	530	
					13,000	= 6.5 ton

MASONRY

Ext. face brick

71–0 × 12–0				=	900 SF	
Deduct						
E1	6–0 × 9–0	=		54		
Wind.	2 × 7–4 × 6–8 ⎫					
	2 × 15–9 × 6–8 ⎬			344		
	2 × 2–9 × 6–8 ⎭					
Sill	51–8 × −5¼			22		
				420	420	
					480	
6.75/SF					× 6.75	
					3,240 pcs.	
+ Cut & waste 6%					195	
					3,435 pcs. =	3.5 M

Wash down face brick = 480 SF

Precast conc. sills 7 × 5¼ (in 6 sills) = 52 LF

8-in. Conc. block backup (8 × 16)   480 SF
    Shelf 71–0 × −8   47
        433
    + ⅛th   54
        487
    + Cut & waste 4%   23
        510 Pcs. 8 × 16

Interior face brick

```
    18-8  ×  9-4   =    174 SF
 less 4-0  ×  3-0         12
                        162 SF
                        ×  6.75
                      1,093 pcs.
      + Cut & waste      57
                      1,150     =  1.15 M
```

12-in. Conc. block walls

```
  Office  51-0  ×   7-0   =    357 SF
  Ext.   425-0  ×  16-0       6,800
  Int.   137-0  ×  14-4       1,964
                              9,121              9,121
  Less:                                   SF
    Drs.         6   ×  18 SF   =    108
    E8        10-0   ×  10-0         100
    E9        16-0   ×  10-0         160
    OH dr.  2 × 10-0 ×  10-0         200
    Int. drs.    3   ×  18 SF         54
    Int. drs.    1   ×  30 SF         30
    Rollup     6-0   ×   4-0          24
    Corridor   6-0   ×   9-0          54
                                     730          730
                                                8,391
               + 1/8 th                         1,049
                                                9,440 pcs.
               + Waste 4%                         380
                                                9,820 pcs.  8 × 16
```

Extra cost for pilasters 16 in. × 4 in. = 4 × 24 c = <u>96 pcs.</u>

Resteel in pilasters
   No. 4 6 × 4 × 18 ft = 432 LF × 0.668 lb = <u>290 lb</u>

Conc. fill to pilasters
   96 Pcs. × 4/9 CF = <u>2 CY</u>

Control joints in 12-in. block walls 8 × 16 ft = <u>128 LF</u>

Build in ends of bar joists = <u>100 ea.</u>

Concrete brick for making good at beams, sills, bar joists, etc., allow <u>1 M</u>

Chapter Eleven

```
Int. conc. block ptns. (8 × 16)
   12 in.    34-0  ×  9-0                    =        306 SF
           -  6-0  ×  8-4  =  50 ⎫
              1 @ 18 SF       18 ⎭          −         68
                                                     238
                              + ⅛th                   30
                                                     268
                              + 5% waste              12
                                                     280 pcs. (8 × 16)

    8 in.    26-0  ×  10-0                            260 SF
            108-0  ×  14-8                          1,584
                                                    1,844
           -  Drs.  6 @ 18 SF    = 108 ⎫
                    6-0 × 4-0      24  ⎬  −          163
                    4-6 × 7-0      31  ⎭
                                                    1,681
                              + ⅛th                   210
                                                    1,891 pcs.
                              + 5% waste              99
                                                    1,990 pcs. (8 × 16)

    6 in.    48-0  ×  8-8                    =        416 SF
            125-0  ×  14-8                          1,834
             89-0  ×  9-4                             831
             19-0  ×  1-8                              32
                                                    3,113
           -  Drs.  8 @ 18 SF    = 144 ⎫
                    4-6 × 9-0      40  ⎭  −          184
                                                    2,929
                              + ⅛th                   366
                                                    3,295 pcs.
                              + 5% waste              165
                                                    3,460 pcs. (8 × 16)

Int. conc. block ptns. (8 × 16) 4 in.
             85-0  ×  9-4  =              794 SF
           -  4-0  ×  3-0         −        12
                                          782
                              + ⅛th        98
                                          880 pcs.
                              + 5% waste   50
                                          930 pcs. (8 × 16)
```

Int. conc. block ptns. (8 × 16) 2 in.
$$
\begin{array}{rl}
19\text{-}0 \times 9\text{-}4 = & 176 \text{ SF} \\
-\ 4\text{-}0 \times 3\text{-}0 & -\ \underline{12} \\
& 164 \\
+\ \tfrac{1}{8}\text{th} & \underline{21} \\
& 185 \\
+\ 5\%\ \text{waste} & \underline{10} \\
& \underline{195}\ \text{pcs.}\ (8 \times 16)
\end{array}
$$

Extra cost 12-in. bond beam blocks
$$
\begin{array}{rl}
3 \times 139\text{-}0 = 417\text{-}0 = & 313 \text{ pcs.} \\
+\ \text{waste} & \underline{7} \\
& \underline{320\ \text{pcs.}}
\end{array}
$$

Reinf. steel to bond beam
  #4   2 × 430 ft = 860 ft @ 0.668 lb = <u>575 lb</u>

Conc. fill to bond beams
  313 pcs. @ 0.5 CF = <u>6 CY</u>

Standard Dur-o-wall reinf. to block walls
  12 in.   8,629 SF @ 1LF/SF = <u>87 CLF</u>
   8 in.   2,114 SF @ 1LF/SF   <u>22 CLF</u>
   6 in.   2,929 SF @ 1LF/SF   <u>30 CLF</u>
   4 in.   782 SF @ 1LF/SF   <u>8 CLF</u>

Masonry scaffold

Ext. brick		900 SF
Ext. block	8 in.	900
Ext. block	12 in.	9,121
Int. block	12 in.	306
Int. block	8 in.	1,844
Int. block	6 in.	3,133
Int. block	4 in.	794
Int. brick		174
		17,172 = 17,180 SF

Mortar (1:3)

Brick	4.65 M @ 18 CF =	84 CF	
12-in. Block	101.0 C @ 10 CF	1,010	
8-in. Block	26.0 C @ 8 CF	208	
6-in., 4-in., 2-in. Blocks	46.0 C @ 6 CF	276	
		1,578	= 59 CY
		say	<u>60 CY</u>

Set loose angle iron lintels 9 × 3 = <u>27 pcs.</u>

Precast concrete lintels

12 in.	×	8 in.	@	4-0 long	=	4 pcs.
8 in.	×	8 in.	@	5-6 long	=	1 pce.
8 in.	×	8 in.	@	4-0 long	=	6 pcs.

CARPENTRY

Wood stud ptn. 2 × 4
    5 × 34 ft = 170 ft
    34 × 10 ft   340 ft
                   510 LF = 340 BF

Roof framing
    2 × 10   38 × 10 ft = 380 ft = 634 BF
    2 × 6   50 ft   50
    2 × 4   50 ft   33
                                            717 = 720 BF

Roof nailers, pressure treated
    2 × 12   2/140 ft + 2/100 ft + 2/10 ft + 52 ft = 552 LF
                                                            = 1,110 BF

Windowsill blocking
    2 × 4   2/8 ft + 2/16 ft + 2/3 ft = 54 LF

1-in. T. & G. roof bdg.
    49-0 × 9-0 = 441 SF
    + cut & waste 20%   89
                             530 BF

1 × 2 Grounds
    Window apron   54 LF
    Wall paneling   486 × 1.2 LF/SF = 582
                                                636 = 640 LF

1 × 3 Grounds
    Base mld.   120 LF

2 × 4 Clg. jsts.
    Pres. toilet 4/6 ft + 2/4 ft = 32 LF

Plaster grounds $\frac{5}{8}$ in. × $\frac{3}{4}$ in.
    Drs. E1., B   3 × 24 ft = 72 LF
    Drs.   13 × 18 ft   234
    Rub. base—Corr.   180
        Off. No. 1   50
        Off. No. 2   70
                                    606 = 610 LF

Temp. closures, windows

$$\left.\begin{array}{rcccc} 2 & \times & 8\text{-}0 & \times & 7\text{-}0 \\ 2 & \times & 16\text{-}0 & \times & 7\text{-}0 \\ 2 & \times & 3\text{-}0 & \times & 7\text{-}0 \end{array}\right\} \quad 54 \times 7 = 378 = \underline{380 \text{ SF}}$$

Temp. closures, doors

6	×	3-0	×	7-0	= 126 SF
		6-0	×	9-0	54
		10-0	×	10-0	100
		16-0	×	10-0	160
					440 SF

Temp. doors   <u>3 opgs.</u>

Rough hardware   <u>L. S.</u>

Set only

H.M. door frame: single, int.	=	<u>20 frames</u>
pair, int.	=	<u>1 frame</u>
H.M. door $1\frac{3}{4}$ in., int.	=	<u>6 doors</u>
12-in. Channel jamb 10 ft high	=	<u>8 pcs.</u>
Ext. H.M. door frame, single	=	<u>6 frames</u>
Ext. H.M. door $1\frac{3}{4}$ in.	=	<u>6 doors</u>
Sliding Kal. fire door 6-6 × 9-6, 2 hr., fus. link, hardware	=	<u>1 ea.</u>

MILLWORK

Solid-core flush wood doors, int.

3-0	×	7-0	×	$1\frac{5}{8}$ in.	=	<u>4 ea.</u>
2-8	×	7-0	×	$1\frac{5}{8}$ in.	=	<u>1 ea.</u>
2-6	×	7-0	×	$1\frac{5}{8}$ in.	=	<u>2 ea.</u>

Hollow-core flush doors, int.

2 @	3-0	×	7-0	×	$1\frac{3}{8}$ in.	=	<u>1 pr.</u>
	3-0	×	7-0	×	$1\frac{3}{8}$ in.	=	<u>1 ea.</u>
	2-8	×	7-0	×	$1\frac{3}{8}$ in.	=	<u>4 ea.</u>
	2-4	×	6-6	×	$1\frac{3}{8}$ in.	=	<u>1 ea.</u>

Overhead wood door & track

10-0	×	10-0	=	<u>3 drs.</u>
16-0	×	10-0	=	<u>1 dr.</u>

Window stool $1\frac{1}{8}$ × 9

    2/8 + 2/16 + 2/3 = <u>54 LF</u> (6 pcs.)

Window apron 1 × 3 = <u>54 LF</u>

1 × 6 Base	+	shoe mold
Pres.	62	
Manager	56	
	<u>118 LF</u>	

Mahogany ply $\frac{1}{4}$ *in. to walls*

Pres.		32–6	×	9–0	⎫	
Manager		21–4	×	9–0	⎬	633 SF
Off. No. 1		16–6	×	9–0	⎭	
		70–4				

— Window			15–9	×	6–8	=	105 ⎫	147
Drs.	2	×	3–0	×	7–0		42 ⎭	
								486 SF

1 × 3 *Trim doors*
2/18 + 2/22 = <u>80 LF</u>

Sliding closet doors $1\frac{3}{8}$ *in. wood incl. track, 4 ft* × *9 ft opg.* = <u>2 pr.</u>

Closet shelf 15 in., 4 ft long, with bearers = <u>2 ea.</u>

Closet hanging rod 4 ft = <u>2 ea.</u>

Chalkboard 4 ft × *3 ft in wood frame* = <u>2 ea.</u>

Toolroom bench 30 in. wide with shelf, 24 in. wide under; 36 in. high = <u>15 LF</u>

Shipping counter 26 ft × *1–6 on 4* × *legs, plus 3 ft* × *3 ft wicket gate* = <u>1 ea.</u>

Shipping wall counter 22 ft × *4 ft* × *3 ft* = <u>1 ea.</u>

Sliding pass window complete, 4 ft × *3 ft* = <u>2 ea.</u>

Set only
    Finish hardware (21 doors)   <u>L. S.</u>
    Toilet accessories   <u>L. S.</u>
    Metal louver 18 × 12, in wood door   <u>3 ea.</u>
    Mat frame 6–0 × 3–0, depressed   <u>1 ea.</u>

ALTERNATES

A-1 Partitions, office No. 2
    Deduct metal office ptns.   <u>Sub</u>

    Add
    2 × 4 stud ptn.

3	×	5	×	8 ft	=	120 LF
		5	×	2 ft		10
3	×	8	×	7 ft		168
		6	×	7 ft		42
						340 LF

$\frac{1}{4}$-in. Birch ply

2	×	3	×	7–0	×	7–0	=	12 sheets 4 ft	×	8 ft
		2	×	2–0	×	7–0		1 sheet 4 ft	×	8 ft
								13 @ 4 ft	×	8 ft
							=	<u>416 SF</u>		

$\frac{3}{4}$ in. × 3 in. Birch base
6/7 ft × 2/2 ft = <u>46 LF</u>

$\frac{3}{4}$ in. × 6 in. Birch cap = <u>24 LF</u>

Return End base 6 in. = <u>4 pcs.</u>

Lacquer finish (painter) <u>Sub</u>

Millwork supply <u>Sub</u>

A-2 Add liquid floor hardener, 2 coats to conc. flrs.
137–0 × 42–0 = 5,754 SF
2 × 54–0 × 43–0 = 4,644
10,398 = <u>10,400 SF</u>

Notes on the take-off (W. D. 9.1 to 9.4)

Although the excavation and site work are shown first, the concrete was actually the first trade taken off. Since many of the excavation items take their dimensions from the concrete quantities, it is prudent and time saving to measure the concrete first. One could do only the concrete below ground, then turn to the excavation, but it is better to complete a trade before leaving it. So the recommended order is—all the concrete, then all the excavation. The excavation and site work are first in the take-off sheet numbering, as in the estimate, because the entire estimate should follow the orderly progression of the project as it is actually built.

This take-off may show a slightly greater tendency to round out computations and final quantities than was found in the prior individual trade examples. This present example is representative of an entire take-off and priced estimate for an actual bid, therefore, it reflects the practical rather than the theoretical. The adjustments are minor and usually round out upward.

Excavation

The stripping is 40 ft around the building, being the area to be graded. The grading area must be cut to 6 in. low to allow for the 6 in. of topsoil for lawns; therefore, 6 in. having already been stripped, the depth to be cut is the same as if from present grade to finish grade. The front projection is taken at the average depth of the SW and SE corners (2–9 + 2–3)÷2 = 2–6. Excavation is now at exterior subgrade throughout, that is, 15 in. below finish floor, but subgrade inside the building should be 11 in. below finish floor, so the correction is made by adding back the building area including ramps, by 4 in. deep.

The wall trench excavation is the same depth as the concrete foundation walls less the average depth of the subgrade below the top of wall. The extra cost for hand excavation is to clean out the bottom of the excavation. It is sometimes measured in square feet as "Trim bottom of footing excavations" and priced at something between 5¢ and 10¢ per square foot.

Note the allowance for compaction in the item listing gravel bed for ground slabs. The shape and trim subgrade item is for the final dozer work to clean up and shape the exterior ready for the topsoil. Topsoil has 20 percent added for compaction because it is very loose material.

Concrete

The wall formwork is measured for full-form panel heights, that is, the 3–4 high wall is taken as 4–0. Premanufactured formwork panels are usually 4–0 wide in multiples of 2–0 high, such as 4–0 × 4–0, 4–0 × 6–0, 4–0 × 8–0, 4 × 10, with filler panels in 6-, 12-, and 18-in. widths. Other odd-size narrow widths and special inside and outside corner pieces are also available.

The concrete in pier footings is a good example of instant conversion to cubic yards—the answer should be apparent immediately when the figures are written (to those who know and apply the "27 times" table). Note that the dock walls, missed when the walls were taken, are taken separately so that the estimator does not have to go back and alter the figures.

The 10 percent for waste and laps added to the wire mesh may be a little heavy. Mesh in flat sheets is easier to handle than in rolls, so 8 percent may be enough to add. The rubbing item is arbitrary, but some would be required to the exposed walls on the outside of the building.

Reinforcing steel

In longitudinal bars an allowance for laps is made at about 1 ft for every 20 ft. Vertical bars are average lengths, since there are no detailed drawings. A small waste allowance, nominally 500 lb, is adjusted upward to 530 lb so as to round out the total tonnage.

Masonry

The face brick is only at the 51–0 × 10–0 front section. The headers are snapped; therefore, there is no extra quantity for headers extending into the backup area, but the cutting and waste allowance is higher than normal. There is some breakage loss when the mason cuts face bricks in half to make snapped headers, so a cutting and waste allowance of 6 percent is carried, rather than the usual 3 to 4 percent. In the block backup item, the brick shelf is deducted. In view of the cutting and fitting of the backup block at the sills, the brick course is not deducted from the concrete block area. The precast concrete sills item is notated "In 6 sills" to facilitate pricing the item.

Concrete block in general—corners are not deducted. This is because of the cutting and fitting of what are 16-in.-long units. In general, door openings are deducted at a nominal area of 18 SF ft per single opening and 30 SF per double opening. These are reasonable quantities for door deductions for any trade, somewhat less than the actual opening sizes, thus leaving some allowance for cutting and fitting without getting involved in tedious separation of the various door sizes. Pilasters and control joints are not described or measured in detail because there are no details available. However, the items must be priced, if only roughly, to get some allowance for them into the estimate.

The concrete brick item is an arbitrary quantity based on experience. Given good details at sills, lintels, and structural steel, the item could be measured reasonably accurately. The bond beam is made up of one course of special U-shaped concrete blocks, open at the ends, filled with concrete, with two reinforcing bars in the bottom. The Dur-o-wall quantity allows for extra courses of reinforcing at the top of walls plus under and over openings. Wall reinforcing in every second concrete block course equals 1 LF of reinforcing per 1.33 SF of wall. Experience has shown that, in actuality, 1 LF per SF is a good average quantity.

The scaffolding is the gross area of walls except that the front section with face brick and backup is scaffolded both sides. The mortar quantities are based on per 1,000 bricks and per 100 pieces of block, respectively. The precast concrete lintels are average length allowing a minimum of 6 in. bearing each end.

Carpentry

To show how the quantity of material for short wood stud partitions is abnormally high, the item is developed in the collection sheet. Horizontal studs are taken as for double sole, double head, one row of bridging. Vertical studs are doubled up at all corners and partition junctions.

Roof boarding 1×6 lays only $5\frac{1}{4}$ in., which is a loss of 14 percent before considering cutting waste.

Where the two go together, millwork items are taken off before completing their corresponding rough carpentry such as grounds, furring, blocking. Temporary closures are taken for all door and window openings, then separately, three operable temporary doors for entrance and egress are provided. The items supplied by subcontractors but set by the carpenter have to be carefully checked when subbids are received to ensure neither overlapping of pricing nor leaving out setting altogether. When in doubt, it is a good rule to take off and price the setting. When the subbids are received, the adjustment is then much easier than rushing to take off and pricing the installation at the last minute. In a big job, the doors would be taken off by types without attempting to separate them by sizes. Job cost records do not segregate setting of wood doors into sizes. They would probably separate interior doors from exterior doors, and wood doors from hollow metal. The wood trim for doors includes the two doorways at wood-panelled walls, plus around the two closets.

Alternates

It is advisable to take off and set up all alternate prices or separate prices while doing the main take-off. The too common practice of leaving these until after the base estimate is all but complete usually results in a rushed, slipshod job because of being pressed for time. Many jobs are won or lost on the alternate prices. Therefore, they should be as carefully prepared as the main bid.

In Alternate #1, note that the plywood for partitions has been measured in the 4 ft × 8 ft sheets which would have to be bought. This example shows the high percentage of waste required for this type of work. Although the net area of plywood required is 322 SF, 416 SF must be purchased to do a proper job without resorting to unsightly and unacceptable patching.

Chapter 12

Industrial Building—The Estimate

The estimate shown on the following pages is for the industrial building in W. D. 9.1 to 9.4 and the take-off in the previous chapter. It represents a typical estimate for a project that a general contractor may submit. The pricing used in the estimate represents a good competitive bid for this particular project, based upon the conditions and location of the job. It is a sample estimate, to demonstrate how a contractor's estimate may be assembled and to explain the reasoning and techniques that were used in the pricing method.

It may be observed that minor adjustments have been made from the take-off to the final estimate. For example, the 2-in. concrete block has been rounded from 195 units to 200 units.

It should be noted that the money extensions have been rounded off. Fractional parts have either been rounded up or down. Many estimators extend their pricing to the nearest dollar. There is nothing wrong with this procedure. The rounding off in this estimate was done for simplification. One will find that the difference in rounding off to the nearest $5 or the nearest dollar will result in very little differences throughout an estimate.

ALTERNATES

It is advisable for the estimator to take off and price all alternate items on separate work sheets. A poor common practice is to wait until the main take-off is complete and then perform an estimate for the alternates. This results, too often, in a rushed and haphazard job that occurs late in the estimating process when time is of a premium. One will find that many jobs are either won or lost based upon the alternate prices. Therefore, a carefully prepared alternate bid should be handled similar to the main estimate.

ESTIMATE—INDUSTRIAL BUILDING

Hawker Machine Inc.
Bids April 8, 19__, 2 P.M.

Sheet	Summary		Labor	Material
1	Job overhead		$19,475	$ 24,520
2	Excavation & site		3,535	14,985
3	Concrete		11,020	9,780
4	Masonry		22,000	10,905
5	Carpentry		3,085	1,165
6	Carpentry millwork		1,365	—
				$ 61,355
			$60,480	60,480
		Total general contractor's work		$121,835
7	Sub-bids			180,985
		Total cost		$302,820
		+ Fee		30,280
	(Bond short $365)			333,100
	Base bid	Base bid		$332,900

Alternate Bid No. 1
Net deduct $ 285

A-1 Deduct $ 285

Alternate Bid No. 2
Net add $1,215
+ Fee   125
               1,340

A-2 Add $ 1,340

INDUSTRIAL BUILDING

No. 1 Job Overhead (General Conditions)		*Labor*	*Material*
Perf. & payment bond		—	$ 2,250
Fire insurance		—	970
Insurance—W. comp., PL, PD } 14 per cent × 60,500		—	8,470
Payroll taxes		—	—
Fringe benefits	in units	—	—
Superintendent	35 wk @ $415	$14,525	—
Field engineers	10 wk @ $325	3,250	—
Job office	8 mo @ $125	—	1,000
Job storage shacks		—	750
Temp. toilets	8 mo @ $ 40	—	320
Temp. telephone	8 mo @ $ 55	—	440
Temp. water		—	200
Temp. light & power		—	850
Temp. heat		700	650
Trucking		150	200
Pickup truck	8 mo @ $120	—	960
Small tools		—	350
Hoisting—forklift	15 wk @ $280	—	4,200
Clean glass		100	—
Clean up		376	50
Premium time	30 wk @ $ 50	—	1,500
Winter protection		175	100
Snow removal		200	—
Sales tax 4 per cent × $31,500		—	1,260
		$19,475	$24,520

INDUSTRIAL BUILDING

No. 2 Excavation & Site		L.	M.	Labor	Material
Strip & stockpile topsoil	750 CY	—	1.20	—	$ 900
Grading exc.	4,540 CY	0.15	1.05	680	4,765
Trench exc. fndtn. walls	435 CY	—	1.30	—	565
Exc. col. ftgs.	60 CY	—	3.00	—	180
Extra—hand exc. ftgs.	18 CY	12.00	—	215	—
Backfill walls & ftgs.					
Bank gravel 95 per cent	270 CY	0.75	3.25	205	880
Site mat'l	211 CY	0.75	1.00	160	210
Hand-trim for ground slab	13,600 SF	0.05	—	680	—
Exc. loading dock	43 CY	0.50	1.25	25	55
Bank gravel bed, dock	14 CY	1.25	3.40	20	50
Bank gravel bed, slab	302 CY	1.25	3.40	380	1,025
Shape & trim site area	25,880 SF	—	0.015	—	390
6 in. Topsoil (site mat'l)	600 CY	0.50	1.85	300	1,110
Lawn seed + lime + fert.	25,880 SF	—	0.06	—	1,555
Disposal surplus (1 mile)	5,600 CY	0.10	0.55	560	3,080
Pumping (surface water only?)	L. S.	100.00	75.00	100	75
Int. trench exc. utilities backfill	95 CY	1.25	1.50	120	145
Exc. platforms	8 CY	11.00	—	90	—
				$3,535	$14,985

INDUSTRIAL BUILDING

No. 3 Concrete		L.	M.	Labor	Material
Formwork, wall ftgs.	1,520 SF	$ 0.55	$ 0.12	$ 835	$ 185
Formwork, fndtn. walls	5,424 SF	0.70	0.12	3,795	650
Formwork, pier ftgs.	180 SF	0.85	0.12	155	20
Formwork, piers	136 SF	1.20	0.12	165	15
Formwork, dock walls	410 SF	0.70	0.12	285	50
Formwork, platforms	105 SF	0.75	0.12	80	15
Brick shelf 8 in.	71 LF	0.80	0.15	55	10
2 × 4 Keyway	760 LF	0.10	0.08	75	60
6-in. Edge form dock	84 LF	0.10	0.35	10	30
Concrete (3,000 lb)					
Wall ftgs.	56 CY	3.00	↑	170	—
Fndtn. walls	93 CY	3.50		325	—
Pier ftgs.	7 CY	6.15		45	—
Piers	2 CY	10.00	see	20	—
Dock walls	5 CY	4.00	below	20	—
5-in. Ground slab	210 CY	3.20		670	—
6-in. Dock slab	7 CY	4.50		30	—
Entrance platforms	4 CY	5.00	↓	20	—
	384 CY		18.50		7,105
Visqueen vapor barrier	15,300 SF	0.02	0.02	305	305
6 in. × 6 in. × 6/6 mesh, flat sheets	15,300 SF	0.06	0.05	920	765
Slab screeds	1,360 LF	0.15	0.05	205	70
Trowel-finish slab	13,590 SF	0.12	0.01	1,630	135
Float-finish dock	360 SF	0.10	—	35	—
Float-finish platforms	160 SF	0.10	—	15	—
½ × 6 Premold exp. jnt.	950 LF	0.12	0.16	115	150
Rub ext. walls	450 SF	0.30	0.06	135	25
1-in. Styrofoam at fndtn. wall	180 SF	0.16	0.18	30	35
Platforms, gravel bed	3 CY	1.50	4.50	5	15
Platforms, rub curb	78 LF	0.30	0.05	25	5
Platforms, 6 in. × 6 in. × 6/6 mesh	170 SF	0.05	0.05	10	10
Reinforcing steel	6.5 T	120.00	Sub	780	—
Reinforcing steel wire & access.	L. S.	—	80.00	—	80
Set col. base plate + 2 A.B.	9 ea.	6.00	5.00	55	45
				$11,020	$9,780

INDUSTRIAL BUILDING

No. 4   Masonry

		L.	M.	Labor	Material
Ext. face brick	3.5 M	$260.00	$87.00	$ 910	$ 305
Int. face brick	1.15 M	260.00	87.00	300	100
Wash down face brick	480 SF	0.12	0.04	60	20
PCC sills 7 × 5¼ (in 6 sills)	52 LF	5.00	5.00	110	110
Conc. block backup (8 × 16)					
8 in.	510 pcs.	0.95	0.34	485	175
Conc. block walls (8 × 16)					
12 in.	9,820 pcs.	1.10	0.42	10,800	4,125
Extra for pilasters					
16 × 4	96 pcs.	1.00	0.50	95	50
Resteel in pilasters	290 lb	0.10	0.10	30	30
Conc. fill pilasters	2 CY	10.00	17.00	20	35
Control joints	128 LF	0.50	0.40	65	50
Build in ends, bar joists	100 ea.	3.00	1.00	300	100
Concrete brick making good at					
sills, beams, etc.	1 M	200.00	55.00	200	55
Conc. block ptns. (8 × 16)					
12 in.	280 pcs.	1.10	0.42	310	120
8 in.	1,990 pcs.	0.95	0.34	1,890	675
6 in.	3,460 pcs.	0.80	0.30	2,770	1,040
4 in.	930 pcs.	0.75	0.26	700	240
2 in.	200 pcs.	0.70	0.24	140	50
Extra for 12-in. bond beam					
block	320 pcs.	—	0.10	—	30
Resteel to bond beam block	575 lb	0.10	0.10	60	60
Conc. fill bond beam block	6 CY	7.00	23.00	40	140
Stand. Dur-o-wall 12 in.	87 CLF	2.00	6.50	175	565
8, 6, & 4 in.	60 CLF	2.00	5.50	120	330
Masonry scaffold	17,180 SF	0.12	0.08	2,060	1,375
Mortar (1:3)	60 CY	3.00	16.50	180	990
Set loose lintels	27 pcs.	4.00	—	110	—
PCC lintels					
12 × 8 @ 4 ft 0 in.	4 pcs.	8.00	15.00	30	60
8 × 8 @ 5 ft 6 in.	1 pc.	6.00	12.50	5	15
8 × 8 @ 4 ft 0 in.	6 pcs.	6.00	10.00	35	60
				$22,000	$10,905

INDUSTRIAL BUILDING

No. 5 Carpentry		*L.*	*M.*	*Labor*	*Material*
Wood stud ptn. 2 × 4	340 BF	$ 0.30	$ 0.15	$ 110	$ 50
Roof frmg. 2 × 10 2 × 6					
2 × 4	720 BF	0.31	0.15	225	110
Roof nailers, pres. treat.	1,110 BF	0.20	0.31	220	345
Roof 1-in. bdg. 1 × 6 T & G	530 BF	0.21	0.16	110	85
Windowsill blocking 2 × 4	54 LF	0.25	0.10	15	5
Ground 1 × 2	640 LF	0.24	0.06	155	40
Ground 1 × 3	120 LF	0.30	0.08	35	95
Ceiling joists 2 × 4	32 LF	0.30	0.10	10	5
Plaster grounds	610 LF	0.25	0.06	155	35
Temp. closures, windows	380 SF	0.25	0.15	95	55
Temp. closures, doors	440 SF	0.30	0.20	130	90
Temp. doors	3 ea.	35.00	25.00	105	75
Rough hardware	L. S.	—	175.00	—	175
Set only					
H.M. doorframe, int.	21 Fr.	18.00	Sub	380	—
H.M. door 1¾ in. int.	6 Dr.	35.00	Sub	210	—
Channel jamb 12 in. @ 10 ft	8 pcs.	20.00	Sub	160	—
Ext. H.M. frame	6 Fr.	20.00	Sub	120	—
Ext. H.M. door 1¾ in.	6 Dr.	35.00	Sub	210	—
Sliding Kal. fire door					
6-6 × 9-6, fus. link, hdwre.	1 ea.	75.00	Sub	75	—
Finish hardware (26 opgs.)	L. S.	360.00	Sub	360	—
Toilet access.	L. S.	140.00	Sub	140	—
Door louver 18 × 12 metal	3 ea.	15.00	Sub	45	—
Mat frame 6 ft × 3 ft	1 ea.	20.00	Sub	20	—
				$3,085	$1,165

INDUSTRIAL BUILDING

No. 6 Millwork		L.	M.	Labor	Material
Set only					
Flush wood door, 1⅝ in. SC	7 ea.	$ 30.00	Sub	$ 210	—
Flush wood door, 1⅜ in. HC	1 pr.	65.00	Sub	65	—
Flush wood door, 1⅜ in. HC	6 ea.	30.00	Sub	180	—
O'head wood door 10 ft × 10 ft	3 dr.	Sub	Sub	—	—
O'head wood door 16 ft × 10 ft	1 dr.	Sub	Sub	—	—
Window stool 1⅛ × 9	54 LF	0.90	Sub	50	—
Window apron 1 × 3	54 LF	0.50	Sub	25	—
Base + shoe mold, 1 × 6	118 LF	0.60	Sub	70	—
Mahogany ply paneling ¼ in.	486 SF	0.38	Sub	185	—
Door trim 1 × 3	80 LF	0.50	Sub	40	—
Sliding closet doors 1⅜ in. wood incl. track, 4 ft × 9 ft opg.	2 pr.	55.00	Sub	110	—
Closet shelf 15 in. × 4 ft	2 ea.	12.50	Sub	25	—
Closet rod 4 ft	2 ea.	8.00	Sub	15	—
Chalkboard 4 ft × 3 ft, wood frame	2 ea.	15.00	Sub	30	—
Toolroom bench 30 in. × 24 in. × 36 in. H.	15 LF	6.00	Sub	90	—
Shipping counter 18 in. wide, 26 ft long on 4 × 4 legs, plus 3 ft × 3 ft gate	1 ea.	120.00	Sub	120	—
Shipping counter 4 ft × 3 ft @ 22 ft long	1 ea.	110.00	Sub	110	—
Sliding glass pass window 4 ft × 3 ft	2 ea.	20.00	Sub	40	—
				$1,365	—

BUILDING

No. 7 Sub-bids

Structural steel	$ 9,840
Open web joists, buy	2,935
Open web joists, install 12.5 T @ $130	1,625
Metal deck	8,400
Misc. iron	1,350
Roofing & sheet metal	10,270
Caulking	400
Millwork } Wood doors	2,355 —
H.M., Kal. drs. & fr.	2,320
Overhead wood drs.	1,370
Alum. windows & entrances } Glass & glazing	4,280 —
Finish hardware allowance	1,650
Roll-up grilles	630
FL, plaster & drywall	1,980
Quarry tile	400
Ceramic tile	2,160
Resilient floors	710
Acoustical material	1,630
Painting	3,300
Toilet access. allowance	275
Metal office ptns.	1,010
Metal toilet ptns.	2,900
Alum. pass windows	170
Door louvers	75
Plumbing	26,800
HVAC	49,750
Electrical	41,000
	179,585
Reinforcing steel to buy	1,400
	$180,985

INDUSTRIAL BUILDING

			L.	*M.*	Labor	Material
No. 8 Alternate Bids						
A-1	*Partitions office No. 2*					
	Add					
	Stud ptns. 2 × 4	340 LF	$0.24	$0.10	$ 80	$ 35
	¼-in. Birch ply	416 SF	0.35	Sub	145	—
	Birch base ¾ × 4	46 LF	0.60	Sub	30	—
	Birch cap ¾ × 6	24 LF	0.55	Sub	15	—
	Base return ends 6 in.	4 pcs.	2.50	Sub	10	—
					$280	35
						280
	Fringe benefits	in units				—
	Ins. & taxes					40
	Sales tax					5
			G. C.'s work			360
	Millwork					205
	Painter					160
				Additions	+	725
	Deduct: metal office ptns.				−	1,010
			Net deductions			$ 285

			L.	*M.*	Labor	Material
A-2	*Floor hardener*					
	Add					
	Liquid floor hardener					
	2 coats to conc. flr.	10,400 SF	$0.075	$0.03	$780	$ 310
						780
						1,090
	Fringe benefits	in units				—
	Ins. & taxes					110
	Sales tax					15
			Net additions			$1,215

NOTES ON THE ESTIMATE

Although the summary sheet and overhead sheet are the first two sheets shown in the estimate, they are the last sheets to be completed. The general contractor's work sheets must be priced before transferring to the summary sheets. In pricing the work sheets, it is prudent to review the contract drawings again, since time will have passed and memories need refreshing. Any work item that is to be performed by the general contractor should be priced early and entered on the summary sheet prior to bid day. Bid day will present enough problems without deliberately creating more difficulties. At the risk of being redundant, we will again express that any and all items that can be priced, checked, and transferred to the summary sheet prior to bid day should be and must be done. A dozen loose items left until the morning of bid day will present unnecessary problems and potentially serious errors. Bid day is hectic enough in getting the sub prices analyzed, completing the last items of the summary sheet and writing up the bid forms.

In general, the procedure should be as follows: (1) complete the general contractor trade sheets and with the labor items from these; (2) compute the overhead items; (3) complete the labor side of the summary sheet and compute the insurance and payroll taxes; (4) transfer the insurance, payroll taxes, health and welfare and similar items to the overhead sheet. Finally, complete the material side of the summary sheet and total all of the general contractor's work items. By the time one is down to pricing the overhead sheet, there should be a good idea of the job cost. Therefore, bond pricing should be completed now. Many estimators will leave the insurance, payroll taxes, bonds, etc., until the last minute in order to get the final overhead items. Although there is nothing wrong with this procedure, there is no justification for delaying as much of the work as can be accomplished early. Any adjustment to the overhead items that may occur later can be covered by a simple adjustment line item to the overhead sheet. This prevents a total oversight of an item being left out, which can occur during the mad rush to get everything together.

The notes that follow are in the order that pricing was accomplished on this proposal. The general contractor performed work items that start with (1) excavation and proceed through millwork, (2) the overhead sheet, (3) subcontractor bids, and (4) summary sheet. The developed unit prices for major items are on pages 222 to 228 and should be referenced to in conjunction with the notes that follow.

Excavation and site

Items that are priced only on the material side of the estimate are representative of rental equipment complete with operator. Pricing of the excavation often depends on the available equipment and the volume of work. In this job, the most efficient equipment for the general grading would have been one or two large self-loading pans. Two 15-CY pans could cut the site in approximately three days at a unit cost of approximately $2.50 to $3.00/CY. This type equipment was not readily available for short-term rental. Presuming this, excavation was priced using a front-end loader and dozer.

In the unit price extensions through the estimate, the labor side covers only the direct field labor. All other items are extended and placed in the material units, including material, equipment, rentals, tools, etc. The lawn seeding was priced on the material side and will probably be subcontracted at a later date either as a unit price or a lump-sum price. Dewatering is included at a nominal amount to take care of any surface water.

Formwork

Note that the formwork is lumped together on the estimate sheet in order to analyze it in its entirety. Since formwork is a carpentry trade item that is separate from the placement of concrete, it can best be handled when reviewing all the work. Further, it provides a better analysis for establishing crews and schedules for performing the work.

Pricing items separately from major formwork items, such as brick shelf, keyways, and edge forms, allows the estimator to establish a unit price for the bulk of the work while pricing the more expensive formwork as "added items." This enables the base unit price to be more consistent with the normal cost of forming and allows good coverage of the more expensive forming items. In doing this, one can rely on historical data from other projects in the current pricing.

Concrete

For simplification and general information, the concrete is totaled and priced as a line item on the material side. The labor unit prices for column footings and piers are high in comparison to the other concrete items. This is due to the small amount of concrete to be placed in nine different locations, causing an expensive operation.

The reinforcing steel was priced for erection only. Purchasing of the steel is a separate subcontract item. However, when the sub sheet was established, the reinforcing steel was missed and was added to the bottom of the sheet. Although out of order, it was not missed.

Masonry

The face brick is included as an allowance of $350.00 per thousand delivered f.o.b. the jobsite. The masonry is entered at $357.00 per thousand in the estimate, the extra $2.00 per thousand was added for breakage during delivery and storage.

Note that all the miscellaneous masonry items such as pilasters, bond beams, control joints, scaffolding, mortar, and minor labor items are priced separately. It is impossible to develop a reasonably accurate unit for labor or material when including mortar, scaffolding, etc. All-inclusive pricing is acceptable for preliminary estimates, but is not good enough for final competitive bidding.

The hoisting equipment for the masonry is not included in the masonry items, but is included in the overhead sheet. It is acceptable to carry hoisting equipment with the masonry work. In this example, however, the hoisting equipment was going to be used for numerous work items. If the equipment is to be used for several trades, it is more appropriate to carry it as an overhead item for the entire project.

Carpentry

Most of the carpentry items were small and the output will be below average. consequently, the labor units are higher than normal.

Items similar to temporary enclosures and rough hardware may be required, but were not included in the estimate. There were no details for setting finish hardware and toilet accessories. Therefore, the labor required to install was included as a lump-sum cost. This pricing was based upon historical information and contractor experience. A rule of thumb for hardware installation costs is approximately 25 to 50 percent of the cost of the hardware material.

Many estimators do not attempt to price the millwork installation. They prefer to use a value of 45 to 60 percent of the subcontractor bid, depending on the nature and quantity of the work. The argument is that the general contractor cannot get enough money for the work and that they always miss some of the items. There may be merit in the argument, but on the surface it appears like running away from one's responsibility. Pricing millwork is like pricing anything else; it starts with a reasonable knowledge of the work involved and a basic idea of the production one person can produce in an hour or day.

Job overhead

The cost of the performance bond was estimated in order to complete the general contractor's work sheets. When the estimate was completed, it was apparent that the initial estimate was low. Therefore, a notation on the summary sheet indicated that the bond was short by $365.

The rates for builder's insurance, payroll insurance, and payroll taxes come from the contractor's records. Fringe benefits were not priced on the overhead sheet in this estimate, because the labor rates used in developing the unit labor price were gross hourly rates. However, a contractor who works in several different areas would best be served by keeping cost records and pricing the labor sheets on net rate of pay, allowing the fringe items to be priced as an additive item on the overhead sheet. Fringe benefits are as much as 25 percent of the net hourly rate for some trades in certain areas, while other areas are lower.

The superintendent, job offices, telephone, and temporary toilet items are based on an anticipated length of time to complete the project. The completion time is often given in the contract documents. In this case, the bidders must indicate the completion time. In large jobs, the contractor may decide to prepare a time schedule to confirm that the work can be accomplished within the specified time or whether premium time must be included in the estimate. By preparing a construction schedule, the general contractor may realize that the work can be performed in a shorter duration than required by the contract documents. This would allow the overhead for the project to be reduced accordingly.

The trucking item is for mobilization, setup, and job clean-up at the completion. Small tools, winter protection, glass cleaning, etc., are all items that are relatively minor but collectively, represent a greater cost.

The sales tax is only on material bought by the general contractor. The subcontractor bid items should be inclusive of sales taxes. All subbids were confirmed to have included sales tax for this project.

Subcontractor bids

For purposes of this estimate, it was considered that all bids were analyzed for completeness, inclusion of sales tax, and in compliance with the design specifications. Subcontractor prices should be solicited, in writing, for their intent to bid the applicable work, with a follow-up telephone call to the major subcontractors. Consideration should be given to solicit bids from contractors that do not have a great deal of work currently on their books. When receiving telephone bids, care must be taken to confirm the bid with a written record of all inclusions, exclusions, variations from the specifications, taxes, total price, alternate prices, unit prices, etc.

As a bit of advice, do not waste time getting numerous prices for an item that is only worth $1,000. Do not make repeated calls to a company that has promised a bid by late morning, leave this item until later and fill in the blanks that are still open.

Mail

If necessary, make arrangements to have mail picked up at the post office rather than waiting for a delivery.

Evaluate the subbids for more than the lowest price, make sure that the bid is complete, decide that the subcontractor will perform the work and not walk away from the job later if there is an error. It is advisable to notify the subcontractor as soon as possible if the bid is considerably low. This gives the subcontractor time to confirm the price to you as well as your competition.

Summary sheet

Having hopefully completed all the general contractor work sheets at least one day prior to the bid, there remains the final computations of the subcontractor sheets and the completion of the summary sheet. All-last minute adjustments should be done on the bottom of the summary sheet. Note that the approximated amount for the bond was short $365 and after making the adjustments the total estimate was $566,914. After the final estimate was determined, there was a management decision to submit the base bid as $566,000 as an enhancement of being the low bidder.

The fee carried in the sample estimate is 10 percent. This was an arbitrary markup selection and is not representative of any particular project. Fees are entirely a question of company policy and need. Considerations of location, job type, competition, risk, work on hand, owner/architect reputations, etc., must be determined on a job-by-job basis before making a final decision.

Alternate bids

The alternates should be set up and priced as far in advance as possible before bid day. Consequently, when the subcontractor bids are received, pricing can be filled in immediately. In the case of relatively minor alternate prices, it is acceptable to include the labor burden for insurance, taxes, benefits, etc., in the alternate. Large alternate prices can be difficult and include multiple subcontractor bids. In this case it will be necessary to require the subcontractor to include alternate prices with their bids. For example:

Electrical base bid		Electrical alternate no. 1	
Sparks, Inc.	$125,000	Sparks, Inc.	+ $23,700
Bright Lights Co.	$130,000	Bright Lights Co.	+ $14,800

Sparks, Inc., can be carried in the base bid, but if Alternate No. 1 is accepted, Bright Light Co.'s total for base plus alternate is $145,400 as compared to Sparks' $148,000. Therefore, the addition for Alternate No. 1 can be included at $20,400. If there were a requirement to name the subcontractor, then Sparks, Inc., should be used for the base and Bright Lights Co. if Alternate No. 1 is accepted.

Chapter 13

Pricing the Estimate

Pricing a general contractor's estimate is not a simple, straightforward task that can be given to an office engineer or junior estimator. A good take-off estimator who is accurate, quick, and responsive to the problems peculiar to the job being estimated may not be capable of pricing the estimate. Pricing requires someone with a working knowledge of what can be accomplished by a crew under certain conditions. An estimator should be a knowledgeable construction person. An estimator must have the ability to visualize a work operation and determine the degree of difficulty in performing the current operation as it compares to past operations by his or her company. Too many estimators spend weeks taking off a job and writing up the estimate without studying the individual work tasks to be performed, and then price the estimate take-off in several hours.

Variations in the final bids by different general contractors can occur for many reasons: (1) casual and haphazard pricing of the estimate, (2) estimators approach the construction task differently, (3) some estimators may have the inside track with an owner or architect or possibly knows the requirements that will ultimately be imposed upon the contractor, (4) some estimators may not be familiar with the existing site conditions or (5) the ability or lack of ability for an estimator to do a good job.

The most dangerous of all bidding methods is the one wherein the estimate is priced "to get the job." This type of contractor, for one reason or another, wants the job so badly that he or she shaves the unit prices, cuts 5 percent off the subtotals (with the intention of buying the cuts out of the subcontractors), and adds little or no fee. The contractor then juggles the total around trying to outguess competitors. This type of contractor is either desperate or is a gambler who will attempt to cut and chisel his or her way out with a profit. Most contractors do not depend on this system or any other guesswork system; they depend ultimately upon good consistent estimating.

THE GOOD ESTIMATOR

Good estimators are those who can evaluate and price almost any job. They recognize and allow for the variables that may occur in the operations and work items. They realize not only the difficulties of a particular job, but also the simplicity that may exist in an operation. One cannot have a mindset that only allows one to see the difficulties, but must realize ways to perform the tasks more efficiently. As stated earlier, good estimators must be knowledgeable construction persons. Good estimators are not expected to know everything and to make all correct decisions, but they must make more right ones than wrong ones. They need to be confident in their abilities, and if confronted with the unknown, must be capable of admitting that they do not have all the answers and be willing to ask questions throughout the industry.

Good estimators do not have a head full of fixed prices ready to throw at the estimate work sheets. If estimating were easy, almost anyone could purchase an existing estimating handbook, open it up, and start pricing. There are numerous estimating books devoted to unit pricing that are extremely useful to an estimator. These can be used to refine the individual's ability to perform his or her tasks. However, the books alone cannot give an individual the necessary experience. There are also many medical books that are available, but without the proper training and experience, it is difficult to diagnose the problem and then perform surgery. The informational books are necessary tools, but not the total answer.

The good estimator is adept at evaluating the subcontractor prices by recognizing incomplete proposals and duplications, and if necessary, can quickly arrive at an estimate to cover obvious omissions. One should not forget that the estimator has a difficult and demanding job and should proceed with caution when estimating unfamiliar projects.

Becoming a good estimator

How does one become a good estimator? The first and foremost requirement is experience. One must have both field and office experience. Observe how an operation is accomplished and what it takes to complete the operation. Keep notes, ask questions, and observe how the more experienced and more successful superintendents get work accomplished. Review the work of the senior estimators and understand their reasoning behind their take-offs and pricing. The development of one's ability to price a contractor's estimate is a gradual and not so obvious process that will take months or years to cultivate and develop. There will be a slow changeover from thinking in terms of money to thinking in terms of output by workers and crews. The ability to see an item as requiring a number of worker-hours per unit of measure is a sign that an estimator will become a good estimator. A competent estimator will always review the job cost records periodically to see how good his or her thought process was during the estimate and periodically visit the job sites to actually see the work being performed. One should make it a point to visit with the project superintendent and discuss the methods that are being used in the field and share thoughts with the field staff.

EXAMPLES OF UNIT PRICES

The examples that follow show how some of the unit prices were determined for the estimate in the previous chapter. The examples demonstrate the methodology used for a specific trade and/or piece of equipment. An estimator should build his or her own unit prices based upon experience.

It should always be remembered that a unit price is an "average" cost for an item. Therefore, it must be based upon good performance as well as bad performance. The examples that follow all refer to the estimate for the industrial building in Chapters 11 and 12. In reviewing the estimate, you may note that the unit prices that were developed will occasionally vary. These variations are by the choice of the estimator in determining what a work activity should cost. Please do not establish a "fixed" unit price in your head and think that you must use it at all times.

General remarks on compiling unit prices

All the examples given of building up unit prices were based on previous experience; they represent the best knowledge that was available at the time of bidding. It could not be guaranteed, however, that the cost records of the completed job would corroborate all the unit prices used. Many factors can affect the actual job cost, and on every job there will be some costs that the estimator did not foresee: some items that he might have anticipated had he been more alert, and some that he could not possibly have expected. A winter that was abnormally severe would play havoc with job costs, yet the estimator could not know that such a spell of terrible weather was coming. A strike at a brick factory hundreds of miles from the site could bring the job to a stop for weeks. Abnormal rainfall might necessitate costly pumping.

Such happenings cannot be foreseen when estimating. Many less unusual cost factors, however, are often missed because the estimate was not awake when he priced the job. There can be no excuse for his not knowing, for example, that a concrete slab over an auditorium is to be 30 ft above the floor and so must be given special consideration when the formwork is priced. Having once had a job in a particular area and so knowing the shortage there of good tradesmen in a particular trade, only a poor estimator would ignore that fact when bidding again in that locality. Having worked for a particular architect before and so knowing that he is extremely fussy about masonry, a good estimator would remember that fact when another of that architect's jobs came up for bidding.

As the estimator receives and dockets such information, it should develop and revise his or her thinking. The wide-awake estimator is constantly enquiring; accumulating, sifting, and evaluating information; learning not to make the same mistake twice.

There will always be something for him to learn about estimating. When an estimator gets the idea that he knows it all, it is safe to say that he is not a good estimator and cannot be without first shaking off that dangerous and restricting conceit.

PRICING THE ESTIMATE

EXCAVATION & SITE WORK

1. *Strip & stockpile topsoil*
   D7 dozer and operator   8 hr × $65.00 = $520.00/day
   Output: 20 CY/hr
   Lost time: 0.75 hr
   Per day: 7.25 hr @ 20 CY/hr   =   145 CY   =   3.58/CY

2. *Grading excavation*
   3 CY shovel dozer and operator   per 8-hr day = $ 440.00
   D7 dozer and operator   per 8-hr day   520.00
   Laborer   8 hr × $10.00   80.00
                                                             1,040.00/day

   Output: 7.25 hr @ 50 CY/hr   =   360 CY   =   2.90/CY
              Labor $0.25/CY   Material $2.65/CY

3. *Disposal 1 mile*
   Front-end loader   8 hr × $50.00   = $ 400.00
   Truck and driver   4 × 8 hr × $42.00   1,344.00
   Laborer   2 × 8 hr × $10.00   1,150.00
                                                     98.40   1,744.00

   9 CY trucks; round trip 15 min
   Trucks: 4 × 8 hr × 4 per hr × 9 CY = 1,150 CY/day
             Labor $0.15/CY   Material $1.50/CY

In Example No. 2 the general contractor's labor was not separated when the total cost, $1,040.00 per day, was developed, but when the unit price was rounded out to $3.58 per cubic yard, the labor was split off. The item shows in the priced estimate at $0.25 labor and $2.65 material.

In Example No. 3 the labor was extended to its separate column for easier computation. This is the preferred way to develop the unit price for ready understanding.

If scrapers were available, Example No. 2 might be as follows:

2A. *Grading excavation*
   Self-propelled 15 CY scraper and
        operator   8 hr × $65.00   = $520.00
   D7 dozer and operator   2 hr × $65.00   130.00
   Laborer   8 hr × $10.00   =   80.00
                                                 80.00   650.00
   Output: 7.25 hr @ 100 CY = 725 CY/day   =   0.11   0.95
   Add: move scraper two ways, 4 hr × $65.00 = $260.00
        $260.00 spread over 4,540 CY   =   0.06
                                                               per CY   0.07   0.56

Note that the cost of moving the scraper in and out is spread over the total yardage in the job.

4. *Trench excavate, building*
    Backhoe ½ CY and operator   8 hr × $48.00 = $152.00
    Output 120 CY/day   = $3.20/CY

5. *Backfill, gravel to buy*
    Bank gravel, delvd. job   ton $2.00
    1.7 T per CY   CY = $3.40   say $3.40
    Shovel dozer and operator   2 hr @ $55.00   = $110.00
    Laborer   2 × 8 hr @ $10.00 $160.00
    Compactor rental per day   2 @ $50.00   100.00
                                                              160.00   210.00
    Material placed: 2 × 8 × 8 CY = 128 CY/day   CY =   1.25   1.64
          Total material price = 3.40 + 1.64 = $3.25/CY

6. *Gravel 6 in., for ground slabs*
    Bank gravel as before   3.40/CY
    Spread—small dozer and operator   1 hr × $55.00 $ 55.00
    Laborer   2 × 8 hr × $10.00 $160.00
    Compactors per day   2 × $50.00   100.00
                                                            160.00   155.00
    Material placed   2 × 8 hr × 5 CY = 80 CY =   2.00   1.93
    Add purchase 1 CY gravel   3.40
                                              Per CY   2.00   5.30

7. *Move, spread, & level topsoil 6 in.*
    Shovel dozer 3 CY & operator   8 hr × $55.00 $440.00
    Dozer D7 & operator   8 hr ×   65.00   520.00
    Laborers   2 × 8 hr ×   10.00 $160.00
                                                          160.00   960.00
    Place 25 CY × 8 = 200 CY/day   per CY =   0.80   4.80

FORMWORK

8. *Formwork material cost*

Wall footings 2 × 12	300 BF	× $300.00 MBF =	$ 90.00
Wall footings 1 × 2	100 LF	× $ 0.10 LF	10.00
Walls 2 × 4/3 × 4	2,800 BF	× $280.00 MBF	784.00
Walls ⅝ in. plywood	900 SF	× $ 50.00 MBF	450.00
Walls 1 × 2	500 LF	× $ 0.10 LF	50.00
Snap ties	600 ea.	× $ 0.15 ea.	90.00
Oil & sundries	allow		160.00
			1,654.00

Total formwork 7,775 SF = 0.21/SF

Above allows: 7 reuses of plywood and lumber
2.5 BF lumber per SF of forms
Wall ties @ 1 per 5 SF of wall surface

9. *Formwork, wall footing*

Carpenters	2 × 8 hr × $12.50 =		$200.00
Laborers	1 hr × $10.00		40.00
	Crew per day		24.00
Output: Erection 160 LF of ftg. =	320 SF =		0.75 SF
Strip: Laborer 2 hr × $10.00 =	$20.00 =		0.06 SF
			0.81 SF

10. *Formwork, foundation walls*

Make up and erect:	Carpenter	2 × 8 × $12.50 =		$200.00
	Laborer	4 × $10.00		40.00
				240.00
Erection: 300 SF/day			=	0.80 SF
Strip: Carpenter 8 × $12.50 =	$100.00			
Laborer 8 × $10.00	80.00			
Output: 750 SF/day	=	180.00	=	0.24
				1.05/SF

The output is for standard panels and does not include the extra labors such as brick shelf and keyways, which are priced at extra cost over the straight wall forms.

11. *Formwork, piers*
    Average pier   1–2 × 1–2 × 3–3 = 15 SF formwork
    Make up: Carpenter   1 hr
    Erect: Carpenter   1 hr
                                    2 hr × $12.50 = $24.00

    Strip: Carpenter $\frac{1}{6}$ hr × $12.50   2.08
    Clean: Laborer $\frac{1}{8}$ hr × $10.00   1.66
                                            15 SF   27.74
                                                    =   1.85 SF

CONCRETE PLACING

12. *Conc., wall footings*
    Small pours at 9 CY per day
    Laborers   3 × 1 hr × $10.00 = $30.00
    Cem. finisher   1 hr × $11.00   11.00
                                     9 CY   41.00
                                         =   4.55 CY

13. *Conc., foundation walls*
    Say 15 CY per hour
    Laborers   3 × 2 hr × $10.00 = $60.00
    Cem. finisher   2 hr × $11.00   22.00
                                     15 CY   82.00
                                        =   5.50 CY

14. *Conc., pier footings*
    4 Ftgs., total 3 CY (borrowed)
    Laborers   2 × 1.5 hr × $10.00   = $30.00
                                                 =   10.00 CY

15. *Conc., ground slab (buggied)*
    A. Base on 2,000 SF per day = 31 CY concrete
         Laborers   2 × 4 hr ⎫
         Laborers   1 × 8 hr ⎬ 16 hr @ $10.00   = $160.00
                                           Placing 31 CY =   3.20 CY
    B. Finishing: Cem. finishers   3 × 8 hr × $11.00   264.00
                    Cem. finishers O.T.   3 × 1 hr × $16.50 =   49.50
                                                    2,000 SF   313.50
                                                            =   0.16 SF

226 Chapter Thirteen

Slab output and rate of pour are based on the average finishing time of 2,000 SF per day for three cement finishers. The overtime is for working through the lunch hour and cleaning up at the end of the day. Note that one laborer is carried for the entire day, as a helper for the finishers.

16. *Reinforcing steel in foundations*

Rodman foreman	8 hr	×	$13.00	=	$104.00
Rodman	8 hr	×	12.00		96.00
Travel expense	2 men	×	6.50		13.00
					213.00

Output per day 2 workers @ 1,250 lb = 2,500 lb = 0.85 per lb
                                              = $170.40 per ton

Masonry
Crew: Masons   8 × 8 hr × 12.50 = $ 800.00
       Laborers   7 × 8 hr × 10.00   560.00
                                      $1,360.00 = $170.00 per mason

Cost per mason day = $111.00

Per worker day
17. *Face brick*   425   = $350.00 per M
18. *Concrete block 12 in.* 100   = $ 1.70 ea
19. *Concrete block 8 in.* 120   = $ 1.42 ea
20. *Concrete block ptns. 6 in.* 140 = $ 1.21 ea
21. *Concrete block ptns. 4 in.* 150 = $ 1.14 ea

Note that the crew has been established at seven laborers to eight masons as an average for the entire job. The output figures take into consideration the fact that the forklift or brick hoist will be carried on the overhead sheet.

22. *Concrete fill to bond beams*

Hi-lift with bucket, rental	2 hr × $25.00	=			$ 50.00
operator	2 hr × 11.00		22.00		
Laborers	2 × 2 hr × 10.00		40.00		
Concrete 3,000 lb	6 CY × 42.00				252.00
	Concrete placed 6 CY		62.00		302.00
	Per CY	=	10.33		42.00

23. *Mortar cost*
    A. 1:3 = Mortar cement   9 bags × $3.00 = $27.00
              Sand 1 CY   = 1.25 ton × 3.00   3.75
                               Material cost per CY = 30.75

    B. Mixing cost
       Laborer (2 bag mixer), 4.5 batches per CY
       Mix 1 batch every 6 min = 27 min/CY @ $6.15 hr = $5.00 per CY

24. *Wash down face brick*
    Mason   8 hr × $13.00 = $104.00
    Laborer   2 hr × 10.00   20.00
         Output 700 SF per day 124.00
                         =   0.18 per SF

CARPENTRY

25. *Stud partitions 2 × 4*
    Carpenter 8 hr × $12.50 = $100.00
    Laborer   0.5 hr × 10.00   5.00
         Output 230 BF/day   105.00
                         =   0.46 per BF

    This allows about 1.5 worker days for the small total amount of wood stud partition. In a job big enough to get fair production, a carpenter should frame 400 to 450 board feet of 2 × 4 studding per day, at a cost of about 0.16 per BF.

26. *Roof nailers, pressure treated*
    Carpenters 2 × 8 hr × $12.50 = $200.00
    Laborer   4 hr × 10.00   40.00
         Output 250 LF, 2 × 12 = 500 BF 240.00
                              =   0.48 per BF

    Note: Pressure-treated material costs more to handle than plain stock; for untreated, use 600 BF per day.

27. *Roof framing 2 × 10*
    Carpenters 2 × 8 hr × $12.50 = $200.00
    Laborers   2 hr × 10.00   20.00
         Output 720 BF   220.00
                              =   0.31 per BF

28. *Roof boarding 1 × 6*
    Total job 530 BF
    Carpenters 2 × 6 hr × $12.50 = $200.00
    Laborers   2 hr × 10.00   20.00
                    530 BF   220.00
                              =   0.42 per BF

29. *Set hollow metal door frame*
    Carpenters 2 × 8 hr × $12.50 = $200.00
    Laborer   2 hr × 10.00   20.00
         8 frames per day   220.00
                              =   25.00 ea.

30. *Set wood door 3 ft × 7ft × $1\frac{5}{8}$ in.*

    Carpenters   2 × 8 hr × $12.50 = $200.00
    Laborers   2 hr × 10.00   20.00
    Laborers unload, stack, and distribute 5 doors: allow 1 hr   10.00
                                        Output 5 doors   250.00

                                                           =   50.00 ea.

31. *Install door hardware*
    Mortice lock set; stop; 3 silencers
        Carpenter   1.5 hr × $12.50 = $18.75 per door
    Add for door closer: 1.2 hr × 12.50 = 15.00 ea.
    Note: installation of butts is included with setting door.

32. *Wood grounds 1 × 2*
    Carpenters   8 hr × $12.50 = $100.00
    Laborer   1 hr × 10.00   10.00
            Output 300 LF/day   110.00

                        =   0.37 per LF

33. *Mahogany paneling $\frac{1}{4}$ in.*
    Carpenters   2 × 8 hr × $12.50 = $200.00
    Laborer   2 hr × 10.00   20.00
      Output 12 sheets 8 ft × 4 ft = 384 SF   220.00

                                     =   0.57 per SF

34. *Stud framing 2 × 4, office ptn.*
    Carpenters   2 × 5 hr × $12.50 = $125.00
                Total job 340 LF =   0.37 per LF
    Note: Unit price is per lineal foot, not per board foot.

35. *Birch plywood $\frac{1}{4}$ in. to office ptn.*
    As in Example No. 33 above 2 carpenters & helper $143.50 per day
              Total job 416 SF = 1 day's work   =   0.53 per SF

36. *Liquid floor hardener, 2 coats*
    1st coat, laborer   800 SF/day   =   8 × $10.00 = $ 80.00
    2d coat, laborer   800 SF/6 hr =   6 × $10.00 = $ 60.00
                                                   800 SF   140.00

                                                                =   0.17 per SF

Appendix

Mensuration and Tables

AREAS

Key to Symbols

L	=	Length	S	=	Number of Sides
W	=	Width	D	=	Diameter
H	=	Perpendicular Height	r	=	Radius
B	=	Length of Base	π	=	3.1416 or $3\frac{1}{7}$

Rectangle: Area = L × W

Parallelogram: Area = L × H

Triangle: Area = $\frac{1}{2}$(B × H)

Regular polygon: Area = S × $\frac{1}{2}$(B × H*)

Circle: Circumference = πD
                 Area = πr^2

Sphere: Surface Area = $4\pi r^2$
               Cubic Volume = $\frac{4}{3}\pi r^3$

Pyramid: Surface Area = B^2 + (2B × Sloping Height)
               Cubic Volume = $\frac{1}{3}B^2$ × H

Cone: Sloping Area = πD × Sloping Height
               Cubic Volume = $\frac{1}{3}\pi r^2$ × H

Cylinder: Curved Surface Area = πD × H
               End Surface Area = $2\pi r^2$
               Cubic Volume = πr^2 × H

*Perpendicular height from side to center point.

LINEAR MEASURE

12	inches	=	1 foot						
3	feet	=	1 yard	=	36 inches				
5.5	yards	=	1 rod	=	16.5 feet				
40	rods	=	1 furlong	=	220 yards	=	660 feet		
8	furlongs	=	1 mile	=	320 rods	=	1,760 yards	=	5,280 feet

SQUARE MEASURE

144	square inches	=	1 square foot		
9	square feet	=	1 square yard		
30.25	square yards	=	1 square rod		
4,840	square yards	=	1 acre	=	43,560 square feet
640	acres	=	1 square mile		

Inches converted to decimals of one foot

1 in.	=	0.083 ft
2		0.167
3		0.250
4		0.333
5		0.417
6		0.500
7		0.583
8		0.667
9		0.750
10		0.833
11		0.917

WEIGHTS OF MATERIALS

Material	*Weight, in lb per CF*
Ashes or cinders	40–45
Bricks (shale commons)	125
Cement	94
Clay (ordinary)	95
Earth (loamy soil)	75–90
Glass	156
Gravel (bank run)	90–105
Iron (cast)	442
Lead	712
Limestone	155–165
Lumber (pine or spruce)	30–32
Marble	160–165
Paper	33–44
Sand — dry	80–85
Sand — wet	90–95
Sandstone	150
Steel	489
Water	62.5
Water (solid ice)	56

CONVERSION TABLES

Lumber

board feet	divided by	12	=	cubic feet
cubic feet	multiplied by	12	=	board feet

Water

cubic feet	multiplied by	62.5	=	pounds
cubic feet	multiplied by	6.25	=	gallons (Imperial)
gallons (Imperial)	multiplied by	1.2	=	gallons (U.S.)
gallons (U.S.)	multiplied by	8.3	=	pounds
gallons (Imperial)	multiplied by	10.0	=	pounds

Temperature

To convert Degrees Fahrenheit to Degrees Centigrade:
  Deduct 32 degrees and multiply by 0.555

To convert Degrees Centigrade to Degrees Fahrenheit:
  Multiply by 1.8 and add 32 degrees

DUODECIMAL MULTIPLICATION

In computing quantities manually, whether because there is no calculating machine at hand or because you simply prefer not to use the machine or slide rule, you will find duodecimal multiplication very useful. Duodecimals are decimals in twelfths, based on the twelve inches per foot.

Example 1

9–8 × 7–11

```
      9– 8
      7–11
     ------
      8–10–4
     67– 8
     ------
     76– 6–4   =   77 SF
```

Explanation

11 in.	×	8 in.	=	88 sq. in.		=		4 (and 7 to carry)
11 in.	×	9 ft	=	99 (plus 7 carried)	= 106	=	8–10	
7 ft	×	8 in.	=	56 twelfths of a SF		=	4– 8	
7 ft	×	9 ft	=	63 SF		=	63	

= 76–6–4

= 76 SF 76 sq. in.

Further Examples

```
12-7  ×  8-5          12- 7
                       8- 5
                      ──────
                       5- 2-11
                      100- 8
                      ────────
                      105-10-11   =   106 SF


9-4  ×  11-3           9-4
                      11-3
                      ──────
                       2-4-0
                     102-8-0
                     ────────
                     105-0-0     =   105 SF


14-3  ×  12-10        14- 3
                      12-10
                      ──────
                      11-10-6
                     171- 0
                     ────────
                     182-10-6    =   183 SF
```

RATIO MULTIPLICATION

It is often quicker to multiply by ratios (or fractions) than to use duodecimals. Ratio multiplication simply applies the ratio between the number of inches and 1 ft by converting the inches to a fraction of a foot. Some of the examples given above would be handled as follows:

```
9-4  ×  11-3          9-4  ×  11 ft   =   102-8
                    + 9-4  ×   ¼      =     2-4
                                           ──────
                                           105-0
                                      =   105 SF


14-3  ×  12-10       12- 0  ×  14 ft  =   168- 0
                   +  0-10  ×  14     =    11- 8
                   + 12-10  ×   ¼     =     3- 2
                                           ──────
                                           182-10
                                      =   183 SF


9-8  ×  7-11          7-11  ×  9 ft   =    71-3
                      7-11  ×  ⅔      =     5-3
                                           ──────
                                            76-6
                                      =    77 SF
```

After some practice it will be found that, between them, duodecimal and ratio multiplication are fast and accurate methods for computing. The ratio method is actually a refinement of the duodecimal; it is simply a quicker way to figure the inches part of a calculation, particularly if one of the figures includes any of the following: 2 in., 3 in., 4 in., 6 in., 8 in., or 9 in. (the less cumbersome fractional parts of a foot).

2 in. = $\frac{1}{8}$ ft
3 in. = $\frac{1}{4}$ ft
4 in. = $\frac{1}{3}$ ft
6 in. = $\frac{1}{2}$ ft
8 in. = $\frac{2}{3}$ ft
9 in. = $\frac{3}{4}$ ft

These manual methods of multiplication should be given careful study. They are really quite simple, and can be mastered very quickly with little effort. Thereafter, they will often be of tremendous help. There are many times when an estimator or an outside engineer, foreman, or superintendent has to compute quantities without being able to turn to a calculating machine. Using these methods and knowing the "27 times" table, you should have no trouble in manually computing quantities. The "27 times" table is given again:

1 × 27 = 27
2 × 27 = 54
3 × 27 = 81
4 × 27 = 108
5 × 27 = 135
6 × 27 = 162
7 × 27 = 189
8 × 27 = 216
9 × 27 = 243

Index

Abrasive aggregate, 80
Alteration work, 151–164
   architectural details of, 151
   demolition, 163
   industrial building, 201–202
   miscellaneous items, 163–164
   occupied buildings, 163–164
   partitions/interior walls, 163
   piping work, excavation/concrete for, 154–156
   problems presented by, 152
   remodeling, 159–163
   roofs adjoining new roofs, 163
   underpinning, 152–154
   wall removal (shoring), 156–159, 163
Anchors, 84
Angle-iron lintels, 120
Arch bricks, 120–121
Areaway and stairs:
   concrete, 61–64
   take-off, 62–64

Backfill, taking-off, 31
Baseball fields, pricing, 51
Beams/suspended slabs, 69–74
   collection sheet for, 70–73
   concrete, 74
      concrete canopy, 74
   sizes, 69–70
Bid bonds, 166
Bid request postcards, 2–3
Blanket insulation, 143
Bonding, masonry, 99
Bonds/insurances, 165–167
   bid bonds, 166
   builder's risk insurance, 166–167
   government payroll taxes, 167
   health and welfare funds, 167

Bonds/insurances (*cont.*):
   pension funds, 167
   permit bonds, 166
Brick anchors, 120
Brick quantities, estimating, 99–104
Brick shelf, 77
Builder's risk insurance, 166–167
Builder's risk insurance, 166–167
Building excavation, examples, 25–29

Cabinets, 149
Calculator, 12
Carpentry, 127–150
   classroom, 133–136
   exterior, 127–132
   finish carpentry, miscellaneous, 149–150
   industrial building, 201, 214–215
   pricing, 227–228
   rough carpentry, miscellaneous, 148–149
   wood-framed buildings
      exterior, 136–143
      interior, 144–148
Casework, 149
Cast-iron pipe, measuring, 49
Catch basins, 41
Ceiling inserts, 84
Cement base, 81
Cesspools, 49
Chain-link fencing, pricing, 51
Chamfer strips, 78
Chimney caps, 80
Cinder running tracks, pricing, 51
Clean-up, as overhead, 170
Clearing site, 23–25
Concrete, curing, 81
Collection sheets, 14–16
   for beams/suspended slabs, 70–73
   for cut and fill, 36

Collection sheets (*cont.*):
  for exterior masonry, 109
  for foundation walls, 15
  for interior masonry, 115
  for partitions, 17
  for roadwork, 37, 38
Color admixtures, 81
Columns, concrete, 67–69
Concrete, 10, 55–84
  areaway and stairs, 61–64
    take-off, 62–64
  beams/suspended slabs, 69–74
    concrete canopy, 74
  columns, 67–69
  equipment pads, 64–67
  foundations, 57–61
  industrial building, 200, 214
  interior stairs, 74–76
  lightweight, yield of, 79
  metal sundries for concrete work, 84
  miscellaneous items, 78–82
  for piping work, 154–156
  precast, 82
  sundry items taken off with, 82–83
  tests, 81
  walks, 80
  wall perimeter, 55–57
  waterproofing admixture for, 81
Concrete-block backup, 103–104, 111
Concrete curbing, pricing, 50–51
Concrete encasing duct, 81
Concrete placing, pricing, 225–227
Construction estimate:
  parts of, 1
  preliminary work, 2–3
  as unique entity, 1–2
Conversion tables, 231
Countertop cutouts, 149
Curbing:
  around curves, measuring, 27
  concrete, 50
    cast-in-place, 50–51
Curbs, 80
Curtain walls, subcontractors' bids, 176–177
Curved boundaries, site, 23, 24
Cut and fill, 33, 36–37
  road, 40

Demolition, 160, 163
Design repetition, 16–17
Dewatering, 50
Distribution box, 29
Door casings, 147
Doors/door frames, subcontractors' bids, 177

Dovetailed anchor slots/anchors, 84
Drain inlets, 49
Drain piping, pricing, 50
Drawings, examination of, 8
Dry wells, 50
Duodecimal multiplication, 231–232

Engineering overhead, 168
Equipment expenses, 169
Equipment pads, 64–67
Estimate, pricing, 219–228
Estimator, 220
Excavation, 21–54
  building, examples, 25–29
  industrial building, 199–200, 213
  for piping work, 154–156
  pricing, 222–223
  rock, 22, 29–32
  for sewer lines, 48
  sheeting for, 31–32
  surplus earth, removal of, 23
  topsoil, stripping, 25
  trench, 41–44
Expansion jointing, 83
Exterior carpentry, 127–132
Exterior masonry, collection sheet for, 109
Exterior scaffolding, 111

Face brick, 97–99, 121, 159
  computing quantities of, 99
Fence postholes, pricing, 51
Fieldstone paving, 50
Fill, transporting to site, 23
Finish carpentry, 149–150
Fire-clay flue lining, 121
Firestop, 142
Flagpole bases, pricing, 51
Flemish bond course, 99–100, 102
Footing drains, 50
Formwork:
  industrial building, 213–214
  pricing, 224–225
Foundations:
  concrete, 57–61
  machine, 80
Foundation walls:
  collection sheet for, 58
  formwork, 77
  perimeter insulation for, 82–83
Four-story building, masonry, 107–112
Framed ground slabs, 77
Framing lumber, 127, 148
Full English bond, 102

Full Flemish bond, 102
Full header bond, 102

Gallery step seating formwork, 78
Glass blocks, 121
Glass breakage, as overhead, 170
Glass cleaning, as overhead, 170
Good estimator, 220
Government payroll taxes, 167
Grade beams, 76
Grading, 32–36
Granolithic floor topping, 81
Gravel bed, taking-off, 31

Hardburned dense face brick, 97–98
Haunches on walls, 77
Headwalls, 49
Health and welfare funds, 167
Heated aggregate, 81
Heavy slabs, 80
Hi-early cement, 81
High slabs, formwork for, 77
Hips, 142, 148

Industrial building:
  alteration work, 201–202
  alternate bids, 216–217
  bids, 182
  carpentry, 201, 214–215
  collection sheet, 187
  concrete, 200, 214
  door schedule, 185–186
  estimate, 203–221
    alternate items, 203
  excavation, 199–200, 213
  finish schedule, 183
  formwork, 213–214
  general contractor's estimate, 179–202
  job overhead, 215
  mail, 216
  masonry, 200–201, 214
  outline specifications, 179–184
    alternate prices, 182–184
    building specialties, 182
    carpentry, 181–182
    concrete, 180
    excavation, 180
    general conditions, 179
    masonry, 180–181
    utilities, 180
  reinforcing steel, 179–202
  site work, 213

Industrial building (cont.):
  subcontractors' bids, 215–216
  summary sheet, 216
  take-off, 184–202
    notes on, 199
Interior masonry, 112–120
  collection sheet for, 115
  details for, 116
Interior scaffolding, 120
Interior stairs, concrete, 74–76
Irregular boundaries, site, 23, 24

Job overhead, 165–172
  bonds/insurances, 165–167
  clean-up, 170
  equipment, 169
  in general, 172
  glass breakage, 170
  glass cleaning, 170
  industrial building, 215
  labor wage increases, 171
  permits, 169
  personnel overhead, 167–168
  premium time (overtime), 171
  progress photographs, 170–171
  sings, 170
  small tools, 169–170
  structures/services, 168–169
  subsistence allowance, 171–172
  travel time, 171
  winter protection, 170

Labor wage increases, as overhead, 171
Lawns:
  sodding, 36
  steel edging for, 50
Lightweight concrete, yield of, 79
Linear measure, 230
Locker bases, 81

Machine foundations, 80
Machine pads, 80
Manholes, 41
Masonry, 97–125
  bonding, 99
  brick quantities, estimating, 99–104
  errors in estimating, 97
  exterior walls with preformed waterproofing, 112
  four-story building, 107–112
  industrial building, 200–201, 214
  interior partitions, 112–120

Masonry (cont.):
  miscellaneous items, 120–122
  preliminary analysis, 98–99
  single-story building, 104–106
    common-brick backup, 106–107
  stonework, 122–125
  waste, 99
Material prices, requests for, 2
Measurements, 229–231
  areas, 229–230
Membrane/special toppings, 82
Metal pan stairs, cement fill for, 80
Metal reinforcing, 120
Metal windows, subcontractors' bids, 176–177
Molded brick courses, 121
Moldings, 149
Mortar, 120, 121–122
Multiplication:
  basic rule for, 12
  duodecimal, 231–232
  ratio, 232–233

Occupied buildings, alteration work, 163–164
Oil-tank manholes/oil-pipe trench, 80–81
Oil-tank mats, 80
Overhead (see Job overhead)
Oversize bricks, 121
Overtime, as overhead, 171

Paneling, 149
Parking-area excavation, 40
Pavings, cut and fill for, 36–37
Pension funds, 167
Perimeter insulation, for foundations walls, 82–83
Permit bonds, 166
Permits, 169
Personnel overhead, 167–168
  engineering, 168
  superintendent, 167
  supervisors, 167–168
  timekeepers/material clerks, 168
  watchman, 168
Pilasters, 77
Plank sheeting, measuring, 32
Planting, 36
Plot plan, 22–23
Plumbing, subcontractors' bids, 176
Polyethylene, 82
Precast concrete, 82
  trench covers, 66

Preformed waterproofing, exterior walls with, 112
Preliminary work, 2–3
Premium time (overtime), as overhead, 171
Pricing the estimate, 219–228
  carpentry, 227–228
  concrete placing, 225–227
  excavation/site work, 222–223
  formwork, 224–225
  good estimator, 220
  unit prices, examples of, 221
Progress photographs, 170–171
Pumping, 49

Quantities, taking off, 8–9
Quantity survey, 7

Ratio multiplication, 232–233
Reinforcing steel, 66, 85–92
  industrial building, 200
  measuring, 54, 86
  take-off:
    for columns, 89
    for foundation, 87–88
    for retaining, 88
    for suspended slab, 90–92
Remodeling, 159–163
Retaining walls:
  pricing, 51
  taking off, 67
Roads and pavings, 36–40
Rock excavation, 22, 29–32
Roof decking, 93
Roof fill, 78–79
Roof rafters, 148
Roofs adjoining new roofs, 163
Rough carpentry, 148–149
Round columns, structural steel, 94, 95

Scaffolding, 159
  exterior, 111
  interior, 120
Seats, pricing, 51
Septic tank, 44–49
  backfill at, 48
  precast concrete, 49
Sewer lines, excavation for, 48
Sewer piping, pricing, 50
Shape and roll, road subbase, 40
Sheeting for excavation, 31–32
Shoring, 156–159, 163

Signs, as overhead, 170
Simpson's Rule of One-Third, 25
Single-story building:
   masonry, 104–106
      common-brick backup, 106–107
Site:
   access to, evaluating, 23
   clearing, 23–25
   deep pit in water, 51–54
   grading, 32–36
   sundry items, 50–51
   visiting, 22–25
Site work:
   industrial building, 213
   pricing, 222–223
Slab bulkheads, 77
Small tools, as overhead, 169–170
Sodding, lawns, 36
Specifications take-off, 3–7
   allowances, 4–7
      carpentry, 6–7
      concrete, 5
      electrical, 7
      excavation and site work, 4–5
      masonry, 5–6
      miscellaneous metal, 6
      structural steel, 5
   alternates, 4
   general conditions, 4
Square measure, 230
Steal edging, pricing, 50
Steel:
   reinforcing, 85–92
   structural, 93–96
Steel joint, taking-off, 96
Step irons, 49
Stonework, 122–125
Structural steel, 93–96
   subcontractors' bids, 174
   take-off for, 94–95
      miscellaneous iron, 96
   troublesome nature of, 93
Structures/services, 168–169
   heat (temporary), 169
   job offices and shanties, 168
   light and power (temporary), 169
   trucking, 168
   water (temporary), 169
Stud framing, 148–149
Stud partitions, measuring, 143
Subbids, requests for, 2
Subcontractors' bids, 173–178
   doors/door frames, 177
   electrical, 174

Subcontractors' bids (*cont.*):
   in general, 176
   glass/glazing, 175
   heating, ventilation, and air conditioning, 174
   industrial building, 215–216
   metal windows/curtain walls, 176–177
   millwork, 175
   miscellaneous items, 175
   miscellaneous steel and iron, 174–175
   plumbing, 176
   requests for, 2
   roofing/flashings, 175
   structural steel, 174
Subflooring, 142
Subsistence allowance, as overhead, 171–172
Superintendent, as overhead, 167
Supervisors, as overhead, 167–168
Surplus earth, removing from site, 23

Take-off:
   collection sheets, 14–16
   general rules for, summary of, 16–19
   mathematical shortcuts, 12–14
   order of, 9–11
   steps prior to, 7–9
   time-saving practices, 11–16
Tarpaulins, 159
Timekeepers/material clerks, as overhead, 168
"Times-ing" method, 16–17
Topsoil:
   determining depth/quality of, 22
   stripping, 25
Travel time, as overhead, 171
Tree wells, 36
Trench excavation, 41–44
Triangulation, site, 23, 24
Trucking, 168
27 times table, 11, 18, 19
   summary of, 16–17

Underpinning, 152–154
Unit prices, examples of, 221
Utilities, 23, 40–51
   collection sheet for, 41
   dewatering, 50
   miscellaneous items, 49–51
      cesspools, 49
      drain inlets, 49
      dry wells, 50
      footing drains, 50

Utilities (*cont.*):
    headwalls, 49
    precast concrete septic tanks, 49
    sewer/drain piping, 50
    septic tanks/distribution boxes, 44–49

Visiting site, 22–25
Visqueen, 82

Walks, concrete, 80
Wall boarding, 149
Wall furring, 132, 135
Wall openings, 77
Wall perimeter, 55–57

Wall perimeter (*cont.*):
    for circular building, 56
Wall removal (shoring), 156–159, 163
Waste, masonry, 99
Watchman, as overhead, 168
Waterproofing admixture, 81
Waterstops, 54, 83
Wedge inserts, for securing steel lintels, 84
Weights of materials, 230
Window blocking, 131–132
Window walls, 149
Winter protection, as overhead, 170
Wood fences, pricing, 51
Wood-framed buildings:
    exterior, 136–143
    interior, 144–148

ABOUT THE AUTHOR

The late NORMAN FOSTER was a leading construction estimator and cost consultant, and the author of *Practical Tables for Building Construction, Construction Pricing and Scheduling Manual,* and other books.

THEODORE J. TRAUNER, JR., P.E., P.P., is principal and CEO of Trauner Consulting Services, Inc., in Philadelphia, Pa., and the author of *Bidding and Managing Government Construction* and *Construction Delays.*

ROCCO R. VESPE, P.E., is director of construction management for Trauner Consulting Services, Inc., with more than 25 years of experience in the management of commercial, industrial, and municipal projects.

WILLIAM M. CHAPMAN is vice president of Trauner Consulting Services, Inc., with more than 25 years of experience in construction management, contracts administration, design-build, estimating, and scheduling.